AF877854

Functional Safety for the Mining and Machinery-based Industries

An integrated framework using AS(IEC)61508, AS(IEC)62061, AS(IEC)61511, ISO13849 and AS4024.1

Marcus Punch

2nd Edition

Cover: Kirstin Farr

2nd Edition

ISBN 978-0-9807660-2-8 (paperback)

RRP $AUD88.00 (including GST), plus postage and handling.

Published by:
Marcus Punch Pty. Ltd.
Tenambit NSW
Australia

Email: marcus@marcuspunch.com
Web: www.marcuspunch.com
Phone: +61 (0)432 168849

Foreword

"Functional Safety for the Mining and Machinery-based Industries" by Marcus Punch is a landmark text.

In risk management and safety literature this book on functional safety fills the void between the various standards and purely technical texts. It encompasses both the practical process and detail necessary to apply functional safety. This is a text that brings together the key standards that alone do not deliver an integrated safety framework.

This is not functional safety simplified, but functional safety clearly explained and referenced for engineers and managers in the mining and machinery-based industries. Detail is not left out nor is the content overwhelming. This is a text as relevant for knowledgeable practitioners as well as for those seeking functional safety information for the first time.

The publication of this book is useful and timely as Governments throughout Australia implement new national occupational health and safety legislation and nationally consistent mine safety legislation.

Marcus clearly explains the purpose of the safety lifecycle to establish safety requirements for plant, considering the specific circumstances and risks in which it will be used, and then to achieve and maintain those requirements for the duration of the plant's life.

The objectives of this book are therefore to provide the reader with:

- an appreciation of the background and reasons why the functional safety and machinery safety approaches should be applied to mining and other plant,
- an overview of the contents of the AS(IEC)61508, AS(IEC)62061, AS(IEC)61511, ISO13849 and AS4024.1 standards, the standards for Functional Safety of Electrical / Electronic/ Programmable Electronic (E/E/PE) safety-related systems and Safety of Machinery,
- an overall framework and stand-alone guideline which integrates and applies these standards to machinery applications in a mine or other facility, and
- an interpretation of the standards, guidance and practical examples

By explaining the integrated safety life-cycle process Marcus guides the reader through the approach to make equipment as safe as possible for people to use. He explores safety integrity level and tolerable risk and the standards that may be applied. All applicable standards are listed and put in context with the working environment of mines and other facilities, the requirement for competent people and safe systems of work to deliver equipment that is fit for purpose throughout its life cycle – the functional safety approach.

Marcus steps the reader through the project life cycle stages of concept, scope, hazard and risk analysis, overall safety requirements, safety requirements allocation, design for operations and maintenance, validation, planning installation and commissioning, safety-related systems, "other technology" safety-related systems, risk reduction facilities, installation and commissioning, operations and maintenance, modification and retrofit and decommissioning.

His style is energetic, clear and knowledgeable.

This book adds significantly to the resources any organisation or individual requires to manage machinery risks and improve safety.

With this book you are equipped to make lasting improvements in the safety of your mining operations.

Rob Regan
Director Mine Safety Operations, NSW
Chief Inspector of Mines and Coal Mines

February 2013.

Table of Contents

List of Tables and Figures

1. Introduction - Why I Wrote This Book.....Again!

"Part of the overall safety relating to the Equipment Under Control (EUC) and the EUC control system that depends on the correct functioning of the Electrical / Electronic / Programmable Electronic (E/E/PE) safety-related systems and other risk reduction measures".[1]

Why did I write this book? Because the subject matter, the terminology and the myriad requirements of the various machinery safety and functional safety standards continue to be a source of confusion for many in industry. Consider for a moment what the prevailing suite of functional safety and machinery standards actually cover...

AS(IEC)61508-2011[2]: Functional Safety of Electrical / Electronic / Programmable Electronic Safety-related Systems

- Consists of eight (8) parts and about 600 pages
- Is based on a sixteen (16) phase 'whole-of-life', or 'cradle-to-grave' safety lifecycle process
- Provides detailed requirements for specifying, designing, validating, operating, modifying and decommissioning electrical / electronic / programmable electronic safety-related systems for all industries
- Uses the Safety Integrity Level (SIL) concept to describe four (4) classes of safety integrity (ie. reliability) – SIL1, 2, 3 and 4. SIL1 describes the lowest level of safety integrity and SIL4 describes the highest. The SIL concept takes account of 'low', 'high' and 'continuous' demand modes of operation of safety-related systems[3].
- The design of "other technology" (ie. mechanical, hydraulic, pneumatic etc...) safety-related systems is not within its scope
- The design of "other risk reduction measures" (ie. guards, barriers, procedures etc...) is not within its scope

AS(IEC)62061-2006: Safety of Machinery – Functional Safety of safety-related Electrical, Electronic and Programmable Electronic Control Systems

- Consists of one (1) part and about 90 pages
- Only covers the 'realisation' (ie. specification, design, installation, testing and commissioning and validation) phases of the 16-phase safety lifecycle process
- Uses the Safety Integrity Level (SIL) concept to describe three (3) classes of safety integrity – SIL1, 2 and 3, as per AS(IEC)61508. However, SIL4 is not within its scope. The SIL concept only takes account of 'high' and 'continuous' demand modes of operation of safety-related systems.
- Refers to AS(IEC)61508 requirements throughout
- Provides detailed requirements for designing electrical / electronic / programmable electronic safety-related systems for machinery-based industries

---

[1] The definition of 'functional safety' provided in AS61508.4.

[2] I have referred to AS61508 and AS62061 throughout this book as AS(IEC)61508 and AS(IEC)62061 for the sole purpose of expressing their origin and equivalence to the IEC standards.

[3] See Appendix A for the AS(IEC)61508.4 definitions for the demand modes.

- The design of "other technology" (ie. mechanical, hydraulic, pneumatic etc...) safety-related systems is not within its scope
- The design of "other risk reduction measures" (ie. guards, barriers, procedures etc...) safety-related systems is not within its scope

AS(IEC)61511-2004: Functional Safety – Safety Instrumented Systems for the Process Industry

- Consists of three (3) parts and about 200 pages
- Is based on the AS(IEC)61508 sixteen (16) phase 'whole-of-life' safety lifecycle process
- Provides detailed requirements for specifying, designing, validating, operating, modifying and decommissioning electrical / electronic / programmable electronic safety-related systems for process-based industries
- Uses the Safety Integrity Level (SIL) concept to describe three (3) classes of safety integrity – SIL1, 2 and 3, as per AS(IEC)61508. However, SIL4 is not within its scope. The SIL concept takes account of 'low', 'high' and 'continuous' demand modes of operation of safety-related systems.
- The design of "other technology" (ie. mechanical, hydraulic, pneumatic etc...) safety-related systems is not within its scope
- The design of "other risk reduction measures" (ie. guards, barriers, procedures etc...) is not within its scope

ISO13849-1:2006 / ISO13849-2:2012 Safety of Machinery: Safety-related Parts of Control Systems

- Consists of two (2) parts and about 200 pages
- Only covers the 'realisation' (ie. specification, design, installation, testing and commissioning and validation) phases of the 16-phase safety lifecycle process
- Uses the Performance Level (PL) concept to describe five (5) classes of safety integrity – PLa, b, c, d and e. PLa describes the lowest level of safety integrity and PLe describes the highest. The PL concept only takes account of 'high' and 'continuous' demand modes of operation of safety-related systems.
- Stated to be compatible with AS(IEC)62061
- Refers to AS(IEC)61508 requirements throughout
- Provides detailed requirements for designing electrical / electronic / programmable electronic safety-related systems for machinery-based industries
- The design of "other technology" (ie. mechanical, hydraulic, pneumatic etc...) safety-related systems is stated to be within its scope
- The design of "other risk reduction measures" (ie. guards, barriers, procedures etc...) safety-related systems is not within its scope

AS4024.1-2006 Safety of Machinery

- Consists of twenty-six (26) parts and about 600 pages.
- Only covers the 'analysis' (ie. hazard identification, risk analysis, safety requirements allocation) and 'realisation' (ie. design, installation, testing and commissioning) phases of the 16-phase safety lifecycle process

- Provides general requirements for specifying and specific requirements for designing and validating "other technology" (ie. mechanical, hydraulic, pneumatic etc...) and electrical / electronic safety-related systems for machinery-based industries
- Uses the Safeguarding Category (CAT) concept to describe five (5) classes of safety integrity – CATB, 1, 2, 3 and 4. CATB describes the lowest level of safety integrity and CAT4 describes the highest. The CAT concept does not classify the demand modes of operation of safety-related systems.
- The design of programmable electronic safety-related systems is not within its scope
- The design of some "other risk reduction measures" (eg. guards, barriers, etc...) safety-related systems is within its scope

Now consider the range of plant that may be encountered in a typical mine or industrial facility...

Machinery-based?	→ Yes[4]
Involves electrical / electronic safety-related systems?	→ Yes
Involves programmable electronic safety-related systems?	→ Yes
Involves "other technology" safety-related systems?	→ Yes
Involves "other risk reduction measures" (eg. guards, barriers etc...)?	→ Yes

Therefore:

- We can't just use AS(IEC)61508 alone. Whilst it provides a 'whole-of-life' safety lifecycle process, it does not cover the design of "other technology" safety-related systems or "other risk reduction measures"
- We can't just use AS(IEC)62061 alone as it does not cover the design of "other technology" safety-related systems or "other risk reduction measures" and it only covers part of the safety lifecycle.
- We can't just use AS(IEC)61511 alone. Whilst it provides a 'whole-of-life' safety lifecycle process, it only covers 'process-like' applications and it does not cover the design of "other risk reduction measures".
- We can't just use ISO13849 alone as it does not cover the design of "other risk reduction measures" and it only covers part of the safety lifecycle
- We can't just use AS4024.1 alone as it does not cover the design of programmable electronic safety-related systems and some "other risk reduction measures" and it only covers part of the safety lifecycle

Confusing?... Frustrating?...

Put simply, none of the above standards are able to provide both a stand-alone safety lifecycle framework and the guidelines necessary for the realisation of a diverse range of safety system applications and technologies that are likely to be encountered in a mine or industrial plant.

[4] But some parts or systems can be process-based, or the machinery is part of a process – See Section 4.3.

Clearly, AS(IEC)61508 provides an appropriate 'whole of life' safety lifecycle process, but what is lacking is a stand-alone guideline which integrates the approaches of AS(IEC)61508, AS(IEC)62061, AS(IEC)61511, ISO13849 and AS4024.1 for the realisation of safety-related systems based on electrical, electronic, programmable electronic and other technologies.

Therefore, this book has been written to draw together the relevant parts of the standards and provide a straight-forward, stand-alone guideline for the application of functional safety in the mining and machinery-based industries.

Whilst the focus is on mining applications, the content is equally applicable to other machinery-based industries.

The objectives of this book are therefore to provide the reader with:

1. An appreciation of the background and reasons why the functional safety / machinery safety approaches should be applied to mining or other plant
2. An overview of the contents of the AS(IEC)61508, AS(IEC)62061, AS(IEC)61511, ISO13849 and AS4024.1, standards
3. An understanding of the key concepts and requirements
4. An overall framework and stand-alone guideline which integrates and applies the AS(IEC)61508, AS(IEC)62061, AS(IEC)61511, ISO13849 and AS4024.1 standards to machinery applications in a mine or other facility
5. Interpretation of the standards, as well as guidance and practical examples

In addition the following resources are provided as appendices to the book:

Appendix A: Glossary of Terms

Appendix B: Project Safety Lifecycle Checklist

2. Why all the Fuss about Functional Safety?

"From 1995 to 2001, there were 11 PE-related (programmable electronics) mining incidents reported in the U.S. and 71 PE-related mining incidents reported in Australia. These incidents are due, in part, to unprecedented levels of system complexity.... Our ability to understand and manage the complexities of PE-based systems has not kept pace with the technology's utilization. As a result, PE-related incidents causing mission failures, harm to the environment, injuries, and fatalities have occurred."[5]

Since the 1980's there has been an increasing use of Programmable Electronic Systems (PES) in safety-related control and protection functions for plant, processes and machinery.

Traditionally, such safety functions were accomplished by mechanical or "hard-wired" electrical systems that had well-understood failure modes and behaviour under fault. The use of programmable electronics and embedded software to perform these safety functions has provided benefits such as flexibility, increased functionality, ease of modification and ease of use. However, the predictability of failure and behaviour under fault is less certain.

A proliferation of PES-related incidents and accidents during the 1980's and 1990's led industries to recognise that it can be very difficult to assure the reliability and behaviour under fault of software and complex electronic hardware. A technical framework was needed to address the use of PES technologies in safety-related applications.

This increasing use of PES gave rise to increasing concerns about the security of these systems. The United Kingdom Health and Safety Executive (UK H&SE) was at the forefront of the response and funded research into the safety integrity of computer systems. In 1987 the UK H&SE issued *"Guidelines for use of PES for Safety Related Applications"*. In this document, the UK H&SE studied accidents involving industrial control systems and classified them according to the lifecycle phase where the causes of the accident occurred. The overall results are shown in Figure 2.1.

Figure 2.1: UK H&SE - Causes of Accidents Involving PES

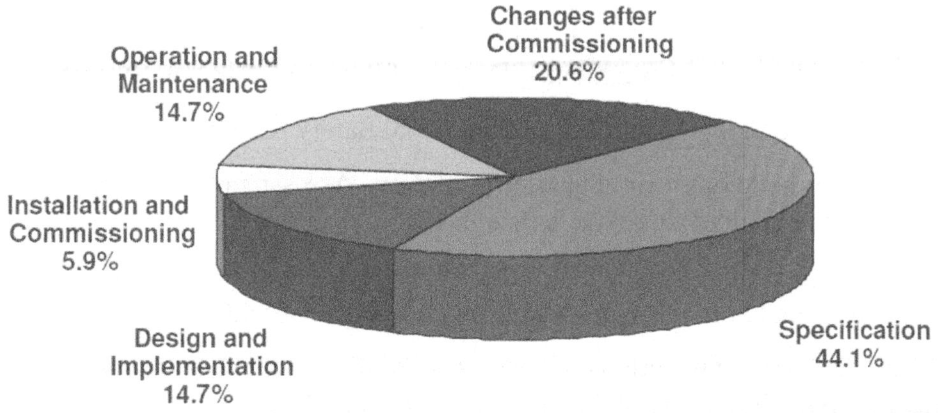

[5] Sammarco., John, J., "*A Complexity Assessment Methodology for Programmable Electronic Mining Systems*", National Institute for Occupational Safety and Health (NIOSH); Pittsburgh, Pennsylvania.

"Guidelines for use of PES for Safety Related Applications" highlighted the need for high levels of rigour at every stage of design, installation and operation of PES. The safety lifecycle concept became formalised in this document - where safety was not just restricted to any particular phase of a system's design process but spanned the whole design process and continued into commissioning, operation/maintenance etc... The safety lifecycle concept evolved to address the accident causes identified in this UK H&SE study.

In Germany, the DIN standards committee developed DIN V19250 with the first edition being released in 1989. This standard introduced the concept of the qualitative 'risk graph' to select one of 8 levels of safety integrity – called the 'AK' class.

In the USA, ISA developed and released the ANSI/ISA S84 standard in 1996. The concept of Safety Integrity Levels (SILs) was firmly established in this standard.

Later, an IEC committee produced the first draft of IEC1508 in 1994. The completed seven (7) part IEC61508 standard was approved between 1998 and 2000. SIL's reached the international arena with the approval of this generic standard for functional safety. Importantly, IEC61508 extended the original scope of PES to include any electrical / electronic / programmable electronic system performing a safety-related control or protective function.

The 7 parts of IEC61508 were adopted by *Standards Australia* as AS61508 during 1999 to 2001. With the inclusion of IEC61508-0 as AS61508-0 in 2006, AS61508 is now an eight (8) part standard.

Parts 1 to 7 of IEC61508 were updated in 2010, and were adopted by *Standards Australia* in 2011.

The IEC61508 standard has since influenced the development of industry-specific and equipment-specific standards for functional safety, including mining and general plant regulations in Australia. These standards and guidelines include (in alphabetical order of their identifier):

- DEF-STD 00-56: Safety Management Requirements for Defence Systems
- DO 178B Air: Software Considerations in Airborne Systems and Equipment Certification
- EES008.1 / EES008.2 / EES008.3 / EES008.4 / EES008.5 (New South Wales Department of Trade and Investment): Design of Powered Winding Systems
- EN12675: Traffic Signal Controllers - Functional Safety Requirements
- EN15233: Methodology for Functional Safety Assessment of Protective Systems for Potentially Explosive Atmospheres
- EN1987-2: Electrically Propelled Road Vehicles - Specific Requirements For Safety - Functional Safety Means And Protection Against Failures
- EN50126: Railway applications - The Specification and Demonstration of Reliability, Availability, Maintainability and Safety (RAMS)
- EN50128: Railway applications - Communications, Signalling and Processing Systems - Software for Railway Control and Protection systems

- EN50129: Railway Applications - Communication, Signaling and Processing systems: Safety-related Electronic Systems for Signaling
- EN50325-5: Industrial Communications Subsystem Based on ISO11898 (CAN) for Controller-device Interfaces - Functional Safety Communication Based on EN50325-4
- EN50402: Electrical Apparatus For The Detection And Measurement Of Combustible Or Toxic Gases Or Vapours Or Of Oxygen - Requirements On The Functional Safety Of Fixed Gas Detection Systems
- EN50491-4-1: General Requirements for Home and Building Electronic Systems (HBES) and Building Automation and Control Systems (BACS) - General Functional Safety Requirements for Products Intended to be Integrated in Building Electronic Systems (HBES) and Building Automation and Control Systems (BACS)
- IEC/EN61000-6-7: Electromagnetic compatibility (EMC) - Generic standards - Immunity requirements for safety-related systems and equipment intended to perform functions in a safety-related system (functional safety) in industrial environments
- IEC/ISO14762: Information Technology – Functional safety requirements for Home and Building Electronic Systems (HBES)
- IEC/TR62685: Industrial communication networks - Profiles - Assessment guideline for safety devices using IEC 61784-3 functional safety communication profiles (FSCPs)
- IEC26262: Road Vehicles – Functional Safety
- IEC60079-29-3: Explosive Atmospheres - Part 29-3: Gas Detectors-Requirements on the Functional Safety of Fixed Gas Detection Systems
- IEC61131-6: Programmable controllers – functional safety
- IEC61326-3: Electrical equipment for measurement, control and laboratory use - EMC requirements - Immunity requirements for safety-related systems and for equipment intended to perform safety-related functions (functional safety)
- IEC61511 (in 3 parts): Functional Safety – Safety Instrumented Systems for the Process Industry. IEC61511 was adopted by *Standards Australia* as AS61511 in 2004
- IEC61513 (in 1 part): Nuclear Power Plants – Instrumentation and Control for Systems Important to Safety – General Requirements for Systems
- IEC61784-3: Industrial communication networks – Profiles – Functional safety fieldbuses
- IEC61800-5-2: Adjustable speed electrical power drive systems – Safety requirements – Functional safety
- IEC62061 (in 1 part): Safety of Machinery – Functional Safety of safety-related Electrical, Electronic and Programmable Electronic Control Systems. IEC62061 was adopted by *Standards Australia* as AS62061 in 2006

- IEC62513: Safety of machinery - Guidelines for the use of communication systems in safety-related applications
- IGEM SR/15: Integrity of Safety-related Systems in the Gas Industry
- ISA/S84.01: Application of Safety Instrumented Systems for the Process Industries
- ISO15998:2008 Earth-moving machinery - Machine-control systems (MCS) using electronic components - Performance criteria and tests for functional safety
- MDG2005 (New South Wales Department of Primary Industries): Electrical Technical Reference for Approval of Powered Winding Systems
- MISRA-C:2004: Guidelines for the Use of the C Language in Critical Systems

Parallel to the development of the functional safety standards, a range of generic "machinery safety" standards have been developed. This suite of standards now includes (but is not limited to):

- ISO12100:2010: Safety of machinery - General principles for design – Risk assessment and risk reduction (content from ISO12100-1:2003 and ISO12100-2:2003 were adopted by *Standards Australia* in AS4024.1201-2006 and AS4024.1202-2006 respectively)[6]
- ISO13849-1:2006 Safety of machinery - Safety-related Parts of Control Systems - Part 1: General Principles for Design (content from EN954-1:1997 and ISO13849-1:1999 was adopted by *Standards Australia* in AS4024.1501-2006)[7]
- ISO13849-2: 2012 Safety of machinery - Safety-related Parts of Control Systems - Part 2: Validation. (content from EN954-2:1999 and ISO13849-2:2003 was adopted by *Standards Australia* in AS4024.1502-2006)[8]
- ISO14121.1:2008 Safety of machinery - Risk Assessment - Principles (content from ISO14121: 1999 was adopted by *Standards Australia* in AS4024.1301-2006)[9]
- ISO14121.2:2012 Safety of machinery - Risk Assessment - Practical guidance and examples of methods

Whilst many similarities exist between the functional safety and machinery safety approaches, there remain fundamental differences which are explained later in this book.

---

[6] ISO12100 was updated and re-published in 2010. Therefore, AS4024.1201-2006 and AS40124.1202-2006 do not reflect the content of the current version of ISO12100.

[7] ISO13849-1 was updated and re-published in October 2006. EN954-1 has been withdrawn. Therefore, AS4024.1501-2006 does not reflect the content of the current version of ISO13849-1.

[8] ISO13849-2 was updated and re-published in October 2012. EN-954-2 has been withdrawn. Therefore, AS4024.1502-2006 does not reflect the content of the current version of ISO13849-2.

[9] ISO14121-1 was updated and re-published in 2008. Therefore, AS4024.1301-2006 does not reflect the content of the current version of ISO14121-1.

3. The Australian Legislative Context

Background

The functional safety approach is fundamentally concerned with the elimination or minimisation of risks posed by processes, plant or machinery hazards to the health and safety of workers and others. It is therefore important that the conduct of a functional safety lifecycle remains consistent with the overarching requirements of work health and safety legislation and regulations in the jurisdictions in which it is undertaken.

Workplace health and safety legislation has recently been reformed in Australia. Previously, each state of Australia has taken responsibility for enacting its own workplace health and safety laws via the provisions of the Australian Constitution. This has contributed to differences in the safety performance achieved within the various jurisdictions as well as inefficiencies and loss of productivity in many organisations which operate in multiple jurisdictions across Australia. The goal of the National Review of Occupational Health and Safety Laws was to develop a nationally consistent Act and Regulation framework that all Australian jurisdictions would accept and adopt.

This harmonisation of Australian work health and safety laws has been part of the Council of Australian Governments (COAG) National Reform Agenda. The *Model Workplace Health and Safety Bill 2011* (the Model Act) is the culmination of a long review process that was commissioned in 2008. The Model Act was accompanied by Model Regulations.

It was intended that all Australian states would enact 'mirror' legislation, adopting the provisions of the Model Act by 1st January 2012. At the time of writing of this book, there were two notable exceptions to this – Western Australia and Victoria have not adopted the Model Act or Model Regulations.

This chapter provides additional context regarding the use of the functional safety approach in Australian jurisdictions where the Model Act and Model Regulations have been adopted. However, the observations and recommendations made later in this chapter regarding risk management processes conducted under AS(IEC)61508 are likely to also apply in the jurisdictions of Western Australia and Victoria where post-Robens[10] style workplace health and safety legislation (which the Model Act is also based on) remains in place.

Specifically, this chapter provides insight into the observed inconsistency between the primary duties under the Model Act and the risk management approaches advocated by AS/NZS ISO31000, AS(IEC)61508 and other standards. Further information on the issues raised in this chapter can be found in the following references:

- Safe Work Australia: *Interpretive Guideline – the Meaning of Reasonably Practicable,* pages 1-7
- Johnstone & Tooma: *Work Health & Safety Regulation in Australia – The Model Act*, pages 94-96, and

---

[10] Lord A Robens, *Report of the Committee on Safety and Health at Work*, 1970-1972, HMSO, London, Cmnd 5034, 1972.

- Johnstone, Bluff & Quinlan: *Submission to the National OH&S Review (Public Submission 055),* pages 21-25

Primary Duties

The Model Act largely follows the architecture of the post-Robens style workplace health and safety statutes that have evolved in the commonwealth and state jurisdictions in Australia. The principal duty holder in the Model Act is the Person Conducting a Business or Undertaking (PCBU). PCBU has been used instead of the term 'employee' and 'worker' has been used instead of 'employee'. This change in terminology has significantly expanded the extent of the duties and obligations under the Model Act.

Clause 19 of the Model Act sets out the primary duty of a Person Conducting a Business or Undertaking (PCBU). The primary duty is concerned with both the health and safety of workers and others. "Health" has been defined broadly in the Model Act to mean physical and psychological health, which includes psycho-social health risks such as stress, fatigue and bullying.

> ***Clause 19 Primary duty of care***
> *(1) A person conducting a business or undertaking must ensure,* ***so far as is reasonably practicable****, the health and safety of:*
> *(a) workers engaged, or caused to be engaged by the person; and*
> *(b) workers whose activities in carrying out work are influenced or directed by the person, while the workers are at work in the business or undertaking.*
> (2) *A person conducting a business or undertaking must ensure,* ***so far as is reasonably practicable****, the health and safety of other persons is not put at risk from work carried out as part of the conduct of the business or undertaking.*

Clauses 20 to 26 of the Model Act set out further duties for certain types of PCBU's that design, manufacture, import, supply, install, construct or commission. For example:

> ***Clause 22 Duties of persons conducting businesses or undertakings that design plant, substances or structures***
> *(2) The designer must ensure,* ***so far as is reasonably practicable****, that the plant, substance or structure is designed to be without risks to the health or safety of persons.*

PCBU's have a duty to protect workers and other persons from harm, but this has been qualified by the concept of "so far as reasonably practicable", or SFAIRP.

Identification and Management of Risks

In support of the PCBU's primary duties, the Model Regulations set out further duties for PCBU's in relation to the identification and management of risks:

> ***34 Duty to identify hazards***
> *A duty holder, in managing risks to health and safety, must identify reasonably foreseeable hazards that could give rise to risks to health and safety.*
> ***35 Managing risks to health and safety***
> *A duty holder, in managing risks to health and safety, must:*
> *(a) eliminate risks to health and safety so far as is reasonably practicable; and*
> *(b) if it is not reasonably practicable to eliminate risks to health and safety—minimise those risks so far as is reasonably practicable.*
> ***36 Hierarchy of control measures***
> *(1) This regulation applies if it is not reasonably practicable for a duty holder to eliminate risks to health and safety.*
> *(2) A duty holder, in minimising risks to health and safety, must implement risk control measures in accordance with this regulation.*
> *(3) The duty holder must minimise risks, so far as is reasonably practicable, by doing 1 or more of the following:*
> *(a) substituting (wholly or partly) the hazard giving rise to the risk with something that gives rise to a lesser risk;*
> *(b) isolating the hazard from any person exposed to it;*
> *(c) implementing engineering controls.*
> *(4) If a risk then remains, the duty holder must minimise the remaining risk, so far as is reasonably practicable, by implementing administrative controls.*
> *(5) If a risk then remains, the duty holder must minimise the remaining risk, so far as is reasonably practicable, by ensuring the provision and use of suitable personal protective equipment.*

Again, the regulations have been qualified by the concept of SFAIRP.

Reasonably Practicable

"Reasonably practicable" has been defined in Clause 18 of the Model Act.

> ***Clause 18 What Is Reasonably Practicable***
> *That which is, or was at a particular time, reasonably able to be done in relation to ensuring health and safety, taking into account and weighing up all relevant matters including:*
> *(a) the* ***likelihood*** *of the hazard or the risk concerned occurring.*
> *(b) the* ***degree of harm*** *that might result from the hazard or the risk.*
> *(c) what the person concerned* ***knows, or ought reasonably to know****, about:*
> *(i) the hazard or the risk, and (ii) ways of eliminating or minimising the risk.*
> *(d) the* ***availability and suitability*** *of ways to eliminate or minimise the risk,*
> *(e)* ***after*** *assessing the extent of the risk and the available ways of eliminating or minimising the risk, the cost associated with available ways of eliminating or minimising the risk, including whether the cost is* ***grossly disproportionate*** *to the risk.*

The Meaning of Reasonably Practicable

Safe Work Australia (SWA) has produced an interpretive guideline to provide further information and clarification on the meaning of "reasonably practicable"[11].

"Risk" is involved in the determination of what is "reasonably practicable" by virtue of the consideration of the likelihood and degree of harm of a hazard. However, it is expected that greater emphasis would be placed on the knowledge of, availability or suitability of ways to eliminate or minimise the risk, particularly for those hazards which have the potential to result in serious injury or death.

The following excerpts from the SWA guideline illustrate the above point:

[11] See http://www.safeworkaustralia.gov.au/sites/swa/about/publications/pages/interpretive-guideline-reasonably-practicable

> *What is 'reasonably practicable' is determined objectively. This means that a duty-holder must meet the standard of behaviour expected of a reasonable person in the duty-holder's position and who is required to comply with the same duty.*
>
> *There are two elements to what is 'reasonably practicable'. A duty-holder must first consider what can be done - that is, what is possible in the circumstances for ensuring health and safety. They must then consider whether it is reasonable, in the circumstances to do all that is possible.*
>
> *This means that what can be done should be done unless it is reasonable in the circumstances for the duty-holder to do something less.*
>
> *If the degree of harm is significant (e.g. death or serious injury is at least moderately likely) then it is unlikely that the cost of implementing available and suitable safety measures to eliminate or minimise the risk would ever be so disproportionate to the risk to justify a decision not to do so.*
>
> *It may be reasonable to expect (and require) a duty-holder to eliminate the risk by ceasing the relevant activity if, after all 'affordable' control measures have been considered, there remains a significant risk of serious injury or illness.*
>
> *The question of what is 'reasonably practicable' is to be determined objectively, and not by reference to the duty-holder's capacity to pay or other particular circumstances. A duty-holder cannot expose people to a lower level of protection simply because it is in a lesser financial position than another duty-holder.*
>
> *If a particular duty holder cannot afford to implement a control that is not so disproportionate to the risk as to be clearly unreasonable, the duty holder should not engage in the activity that gives rise to that hazard or risk.*

In summary, it would appear that a large part of satisfying the "reasonably practicable" test relies on the PCBU adopting all of the safety measures for a hazard that are known, available and suitable at the particular time. If the degree of harm is significant (ie. serious injury or death) then any "gross disproportionality" argument discounting the use of a known, available and suitable safety measure is unlikely to be justifiable. Furthermore, if a PCBU cannot afford a known, available and suitable safety measure, and "gross disproportionality" cannot be justified, then the activity that gives rise to the hazard should not be undertaken.

Is the Use of Functional Safety Standards "Reasonably Practicable"?

AS(IEC)61508, AS(IEC)62061, AS(IEC)61511, ISO13849, AS4024.1 and the other standards mentioned in this book provide processes, methodologies, techniques and measures to ensure that E/E/PE safety-related systems and "other technology" safety related systems are adequately specified, designed, verified, installed and commissioned, validated, operated maintained, modified and decommissioned.

Standards are part of what is "known" about a subject. PCBU's are not obliged to comply with standards however they are required to identify relevant standards and consider them. A court may have regard to standards as evidence of what is, or was, known about a hazard, risk, risk assessment or risk control.

The above-mentioned standards are part of what is "known" about how to determine what safety-related systems are required in a plant to reduce risk, how to design them be reliable and how to effectively manage them throughout the life of the plant.

Therefore, the use of these standards would contribute to a PCBU's overall argument that either:

- The health and safety of workers and others at a particular place of work had been ensured "so far as is reasonably practicable", or
- The plant had been designed to be without risks to persons "so far as is reasonably practicable".

It may be difficult for a PCBU to argue that the use of these standards was not "reasonably practicable" if others in a similar situation had used them effectively.

Aligning the Functional Safety Approach with the Model Act and Regulations

The PCBU's primary duties and the regulations pertaining to risk management have significant implications for the conduct of the functional safety approach within those Australian jurisdictions which have adopted the Model Act and Model Regulations, as well those jurisdictions that have not.

In particular, the conduct of Phase 3 (Hazard and Risk Analysis), Phase 4 (Overall Safety Requirements) and Phase 5 (Overall Safety Requirements Allocation) are affected. This can be explained by referring to the general concept for risk reduction described in AS(IEC)61508.5 Figure A.1 (shown here in Figure 3.1).

Figure 3.1: AS(IEC)61508 General Concept for Risk Reduction

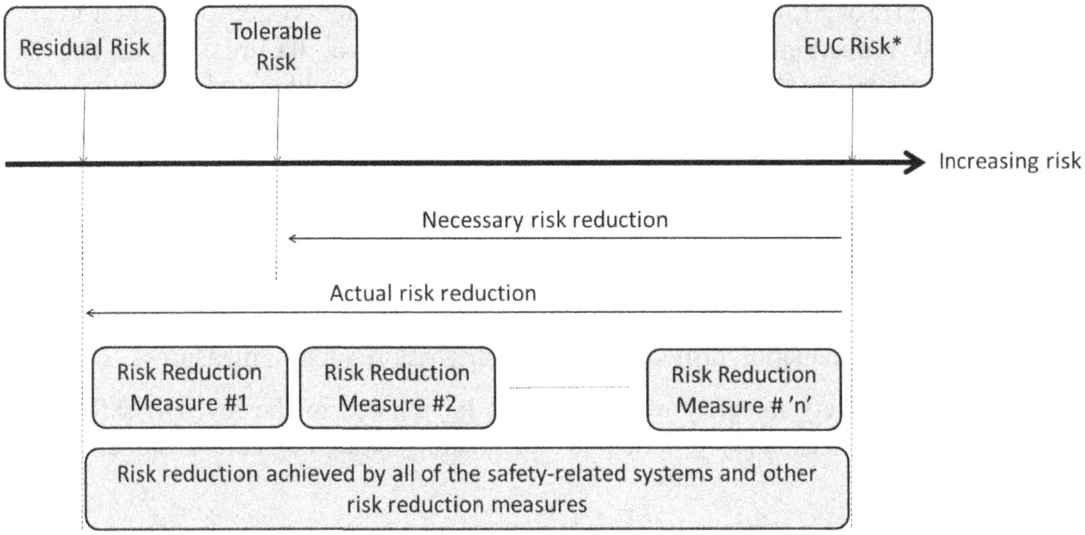

EUC Risk: a risk arising from the equipment Under Control (EUC) or its interaction with the EUC control system.

Phases 3 to 5 of the AS(IEC)61508 functional safety lifecycle follow a "tolerable risk" approach to risk management that is consistent with the model of risk management advocated in the AS/NZS ISO31000 Risk Management standard.

For each EUC risk that is identified, the level of risk is estimated and then one or more risk reduction measures are selected. In AS(IEC)61508 risk reduction measures are broadly divided into two categories: safety-related systems and "other risk reduction measures". Each risk reduction measure provides its own degree of risk reduction against the EUC risk. Collectively, all of the risk reduction measures provide an overall degree of risk reduction against the EUC risk, known as the "actual risk reduction". As the EUC risk has not been eliminated, a level of "residual risk" remains in place, following the implementation of all of the risk reduction measures.

The aim of this approach to risk management is to apply sufficient risk reduction measures against the EUC risk such that the "actual risk reduction" exceeds the minimum level of risk reduction that would be needed (ie. the "necessary risk reduction") to achieve a notional level of "tolerable risk" which is set by the end-user of the EUC. The notion of "tolerable risk" is discussed in detail in Section 7.5.

The AS(IEC)61508 general concept of risk reduction is therefore at odds with the PCBU's primary obligation (Clause 19) under the Model Act.

Under the Model Act, the PCBU's obligation is <u>not</u> concerned with 'tolerable risk'. Instead it is focused on ensuring a level of safety that would allow the test of "reasonably practicable" to be met. Practically, this may mean that all known, available and suitable risk reduction measures for a hazard would need to be put into place, unless it was reasonable not to.

Furthermore, AS(IEC)61508.1 Clause 7.4.2.2 states in regard to hazard and risk analysis phase that *"consideration shall be given to the elimination or reduction of the hazards"*. This is a much less onerous requirement than that set out by the Model Regulations Clauses 34-36, which requires that all reasonably practicable measures to eliminate a hazard must be considered before measures that reduce the risk of the hazard are considered.

Furthermore, during the safety requirements allocation phase the level of safety integrity (ie. the SIL or PL or CAT) allocated to a safety-related system is related to the degree of risk reduction afforded by that system against the hazards posed. The level of risk reduction (and therefore the level of safety integrity) allocated to a safety-related system will be dependent on a calculus of the factors shown in Figure 3.1 (SIL allocation is covered in detail in sections 7.6 and 7.7).

- The level of EUC risk
- The "tolerable risk" target
- The number of "other risk reduction measures" put in place against the EUC risk
- The collective risk reduction provided by the "other risk reduction measures"

This means that the safety integrity allocations are not related to the level of safety integrity that may be 'reasonably practicable' for a safety-related system designer to achieve.

The AS(IEC)61508 general concept for SIL allocation is therefore at odds with the designer PCBU's obligation under Clause 22 of the Model Act.

Clause 22 would have the designer target a level of safety integrity that is "reasonably practicable" to achieve, not one based on the achievement of a notional level of "tolerable risk".

For example, it is possible that a SIL allocation process based on the general concept for risk reduction may result in a SIL1 allocation being made to a particular machinery safety function, such as an Emergency Stop. However, on the basis of what is known, available and suitable it may actually be "reasonably practicable" to design the Emergency Stop to higher level of safety integrity (such as SIL2 or even SIL3) without a "grossly disproportionate" increase in the cost of implementation.

I therefore recommend that the following actions are taken to ensure that any functional safety approach (and particularly Phases 3 to 5) that is undertaken is consistent with the requirements of the Model Act and Regulations:

1. Risk elimination measures must be used against a hazard, where "reasonably practicable", before risk reduction measures are considered.
2. If risk elimination measures are not "reasonably practicable", the hierarchy of risk controls must be used to guide the selection of the most effective risk reduction measures for each hazard.
3. The selection of risk reduction measures for each hazard must encompass all those that are known, available and suitable at the particular time. All that can be done should be done.
4. If the potential severity of harm from the hazard is significant (ie. serious injury or death) then all known, available and suitable risk reduction measures should be implemented, regardless of cost, or the activity giving rise to the hazard should not be undertaken.
5. Safety requirements allocation against the hazard using a 'tolerable risk' approach should not occur until there is a level of confidence that the selection of risk reduction measures to be used against the hazard would allow the "reasonably practicable" test of the PCBU's primary duty to be satisfied.

6. The allocation of levels of safety integrity (ie. SIL, PL or CAT) to any safety-related systems should also take account of the level of safety integrity that would be "reasonably practicable" for a designer to achieve.
7. Where a regulation, code of practice or standard specifies a level of safety integrity or other specific requirements for a particular type of safety-related system, and this is more onerous than the SIL, PL or CAT allocated, then this requirement should take precedence.

4. An Overview of the Standards

The following standards are those that are most relevant to functional safety / machinery safety of mining and industrial plant and machinery:

- AS(IEC)61508 Parts 0 to 7: Functional Safety of electrical / electronic / programmable electronic safety-related systems.
- AS(IEC)62061 Safety of machinery - Functional safety of safety-related electrical / electronic / programmable electronic control systems.
- AS(IEC)61511: Functional safety – Safety instrumented systems for the process industry sector
- ISO13849: Safety of machinery - Safety-related parts of control systems
- AS4024.1 Safety of machinery

Other equipment-specific standards such as AS4240 (Remote Controls for Mining Equipment), AS4024.3 Parts 3610 to 3615 (Safety of Machinery - Conveyors, superceding AS1755:2000 Conveyors – Safety Requirements) and AS1418.10:2011 (Cranes, Hoists and Winches – Mobile Elevating Work Platforms) also refer to AS(IEC)61508, AS(IEC)62061, ISO13849 or AS4024.1.

A glossary of the terms and definitions used by these standards is provided in Appendix A.

4.1 AS(IEC)61508

This standard originated from IEC as IEC61508. IEC61508 now has eight (8) parts. Parts 1, 2, 4 & 5 were originally issued in December 1998, Parts 3, 6, 7 were originally issued in May 2000 and Part 0 was originally issued in 2005. All 8 parts of the standard were adopted by Standards Australia as AS61508 during 1999, 2001 and 2006 respectively. Parts 1 to 7 of IEC61508 were updated in 2010, and were adopted by *Standards Australia* in 2011.

AS(IEC)61508 sets out a generic approach for all safety lifecycle activities for systems comprised of electrical and/or electronic and/or programmable electronic systems. The standard is not prescriptive in nature. It adopts a 'risk-based' approach for the determination of 'performance-based' safety requirements.

The Safety Integrity Level (SIL) concept is used to describe four (4) classes of safety integrity (ie. reliability) – SIL1, 2, 3 and 4. SIL1 describes the lowest level of safety integrity and SIL4 describes the highest. The SIL concept takes account of 'low', 'high' and 'continuous' demand modes of operation of safety-related systems[12].

For 'low' demand safety functions, safety integrity is measured by Probability of Dangerous Failure on Demand (PFD). The maximum PFD's allowable for each level of safety integrity are described as follows in AS(IEC)61508.1 Table 2.:

SIL1: PFD < 0.1 (or 10^{-1})

SIL2: PFD < 0.01 (or 10^{-2})

SIL3: PFD < 0.001 (or 10^{-3})

SIL4: PFD < 0.0001 (or 10^{-4})

[12] See Appendix A for the AS(IEC)61508.4 definitions for the demand modes.

For 'high' and 'continuous' demand safety functions, safety integrity is measured by Probability of Dangerous Failure Per Hour (PFH). The maximum PFH's allowable for each level of safety integrity are described as follows in AS(IEC)61508.1 Table 3:

SIL1: PFH < 0.00001 (or 10^{-5})

SIL2: PFH < 0.000001 (or 10^{-6})

SIL3: PFH < 0.0000001 (or 10^{-7})

SIL4: PFH < 0.00000001 (or 10^{-8})

The 8 parts to the standard are:

Part 0: Overview of Functional Safety and AS(IEC)61508

Part 1: General Requirements

Part 2: Requirements for Electrical / Electronic / Programmable Electronic Safety-related Systems

Part 3: Software Requirements

Part 4: Definitions and Abbreviations

Part 5: Examples of Methods for the Determination of Safety Integrity Levels

Part 6: Guidelines on the Application of AS(IEC)61508.2 and AS(IEC)61508.3

Part 7: Overview of Techniques and Measures

Each part is described below in more detail.

Part 0: Provides a summary of the concept of functional safety and an overview of the AS(IEC)61508 series of standards.

Part 1: Provides the overall safety lifecycle framework for the achievement of functional safety. It covers the overall requirements for:

- Documentation of the safety lifecycle process (Clause 5 and Annex A)
- Management of functional safety, including competence of persons (Clause 6)
- The overall safety lifecycle process (Clause 7.1)
- The scoping, hazard and risk analyses leading to the specification of safety requirements (Clauses 7.2 to 7.6 and 7.10)
- Design, installation, commissioning and validation of the safety-related systems, including the planning requirements for these activities (Clauses 7.7 to 7.9 and 7.11 to 7.14)
- Operation and maintenance, modification and retrofit, decommissioning or disposal of safety-related systems (Clause 7.15 to 7.17)
- Verification (Clause 7.18)
- Independent functional safety assessment (Clause 8)

Part 2: Provides specific requirements for the various activities associated with the realisation of electrical / electronic / programmable electronic safety-related systems (except for software – covered in Part 3), including:

- Safety Requirements Specification (Clause 7.2)
- Safety Validation Planning (Clause 7.3)
- Design and Development (Clause 7.4)

- Integration (Clause 7.5)
- Operation and Maintenance Procedures (Clause 7.6)
- Safety Validation (Clause 7.7)
- Modification (Clause 7.8)
- Verification (Clause 7.9)
- Techniques and measures for control of failures during operation (Annex A)
- Techniques and measures for avoidance of systematic failures during different phases of the lifecycle (Annex B)
- Techniques for the calculation of Diagnostic Coverage and Safe Failure Fraction (Annex C)
- Contents of a safety manual for compliant items (Annex D)
- Special architecture requirements for integrated circuits (IC's) with on-chip redundancy (Annex E)
- Techniques and measures for ASIC's (application-specific integrated circuits) – avoidance of systematic failures (Annex F)

Part 3: Provides specific requirements for the various activities associated with the realisation of safety-related systems software, including:

- Additional requirements for management of safety-related software (Clause 6)
- Software Safety Requirements Specification (Clause 7.2)
- Software Safety Validation Planning (Clause 7.3)
- Software Design and Development (Clause 7.4)
- Programmable Electronics Integration (Clause 7.5)
- Software Operation and Maintenance Procedures (Clause 7.6)
- Software Safety Validation (Clause 7.7)
- Software Modification (Clause 7.8)
- Software Verification (Clause 7.9)
- Selection of techniques and measures (Annexes A and B)

Part 4: This part covers all of the definitions and abbreviations used in the other parts.

Part 5: This part covers the various risk-based approaches for the development of the Safety Integrity Level (SIL) requirements, including:

- The underlying concept of risk reduction and the relationship of risk to safety integrity (Annex A)
- The selection of methods for SIL determination (Annex B)
- The underlying concepts of tolerable risk and the ALARP principle (Annex C)
- A quantitative method for SIL determination (Annex D)
- Risk graph methods for SIL determination (Annex E)
- Layer of Protection Analysis (LOPA) method for SIL determination (Annex F)
- Hazardous event severity matrix method for SIL determination (Annex G)

Part 6: This part contains guidelines for the application of Parts 2 & 3, including:

- A brief overview of the requirements of Parts 2 & 3, setting out the functional steps in their application (Annex A)
- A worked example of the technique for calculating the probabilities of hardware failure (Annex B)
- A worked example of the technique for calculating Diagnostic Coverage and Safe Failure Fraction (Annex C)
- A methodology for quantifying the effect of hardware-related common cause failures (Annex D)
- Examples of the application of the software safety integrity tables specified in Part 3 (Annex E)

Part 7: This part presents an overview of the techniques and measures relevant to Parts 2 and 3, including:

- Control of random hardware failures - such as monitoring of relay contacts, voting architectures, de-energise to trip, separation of electrical and information cabling, etc...(Annex A)
- Avoidance of systematic faults - such as project management, traceable documentation, skilled operators, etc...(Annex B)
- Achieving software safety integrity - such as data flow diagrams, temporal logic, defensive programming, on-line checking, certified tools, use of verified modules, graceful degradation, simulation etc...(Annex C)
- A probabilistic approach to achieving software safety integrity for pre-developed software (Annex D)
- An overview of the techniques and measures for the design of ASIC's (Annex E)
- Definitions of the properties (ie. desired outcomes) of the software lifecycle phases (Annex F)
- Guidelines for the development of safety-related object-oriented software (Annex G)

4.2 AS(IEC)62061

This standard originated from IEC as IEC62061. IEC62061 has one part. It is machine-sector specific whilst maintaining the functional safety framework of IEC 61508. Reference is made to IEC61508 throughout the standard. IEC62061 was released in 2005 and adopted as AS62061 in 2006.

In December 2009, IEC62061 (and ISO13849) were adopted as harmonised European Standards under Machinery Directive 2006/42/EC.

AS(IEC)62061 is intended for use by end-users, machinery designers, control system manufacturers and integrators, and others involved in the design and development of Safety-related Electrical Control Systems (SRECS).

The Safety Integrity Level (SIL) concept is used to describe three (3) classes of safety integrity – SIL1, 2 and 3, as per AS(IEC)61508. However, SIL4 is not within its scope. The SIL concept only takes account of 'high' and 'continuous' demand modes of operation of safety-related systems. The maximum PFH values for each SIL are the same as those described in AS(IEC)61508 Table 3 (see Section 4.1).

It sets out an overall design and development approach and specific requirements to achieve the necessary safety integrity, including:

- Management of functional safety (Clause 4)
- Specification (Clause 5)
- Design and integration (Clause 6)
- Information for the use of the safety-related systems (Clause 7)
- Validation (Clause 8)
- Modification (Clause 9)
- Documentation (Clause 10)

Guidelines are also provided for machinery-sector users, including:

- A generic machine-sector methodology for Safety Integrity Level (SIL) assignment (Annex A)
- A design guideline for the development of safety-related systems (Annex B)
- A guideline to the design and development of embedded software (Annex C)
- A list of failure modes for general electrical / electronic equipment (Annex D)
- Guideline on the application of IEC61000 electromagnetic immunity standards to safety-related systems (Annex E)
- Methodology for the estimation of susceptibility to common cause failures (Annex F)

Importantly, equivalence between the AS4024.1 machinery safeguarding categories (CAT B, 1, 2, 3 and 4) and AS(IEC)61508 / AS(IEC)62061 Safety Integrity Levels (SIL 1, 2 and 3) has been drawn in AS(IEC)62061-2006 Table 6. This table implies that a safety function designed and validated to a particular AS4024.1 CAT may be deemed to be of an equivalent level of safety integrity to one designed to an AS(IEC)61508 / AS(IEC)62061 SIL.

However, this equivalence is now outdated due to the updating of ISO13849 that has occurred since 2006 and the replacement of Safeguarding Categories (CAT's) with Performance Levels (PL's) as the safety integrity target measure in that standard[13].

A joint IEC / ISO technical report has been prepared for users of ISO13849 and IEC62061[14]. It explains the relationship between the two standards and the equivalence between PL and SIL. ISO and IEC have both stated that at some point in the future IEC62061 and ISO13849 will be merged into one standard.

---

[13] See Section 4.4 concerning the validity of AS4024.1 design and development approach in the light of changes to ISO13849.

[14] Issued jointly in July 2010 as IEC/TR 62061–1 and ISO/TR 23849. These were produced by a Joint Working Group (JWG) consisting of representatives of ISO/TC199 and IEC/TC44.

4.3 AS(IEC)61511

This standard originated from IEC as IEC61511. IEC61511 has three (3) parts, all originally issued in 2003. All 3 parts of the standard were adopted by Standards Australia as AS IEC 61511 in 2004. IEC61511 was in the process of being updated by IEC at the time of writing of this book. It is anticipated that IEC61511 may be re-issued in 2013 or 2014 with a re-issue of AS IEC 61511 thereafter.

This standard was developed as a process industry application of parts of AS(IEC)61508 in the same manner that AS(IEC)62061 was developed as a machinery industry application of parts of AS(IEC)61508.

The AS(IEC)61511 standard provides requirements for the specification, design, installation, operation and maintenance of Safety Instrumented Systems (SIS) that can be "confidently entrusted to place and / or maintain a process in a safe state"[15]. Like AS(IEC)61508 it sets out a generic approach for all safety lifecycle activities for systems comprised of electrical and/or electronic and/or programmable electronic systems.

The Safety Integrity Level (SIL) concept is used to describe three (3) classes of safety integrity – SIL1, 2 and 3, as per AS(IEC)61508. However, SIL4 is not within its scope. The SIL concept takes account of 'low', 'high' and 'continuous' demand modes of operation of safety-related systems. The maximum PFD and PFH values for each SIL are the same as those described in AS(IEC)61508 Tables 2 and 3 (see Section 4.1).

AS(IEC)61511-1 sets out an overall design and development approach and specific requirements to achieve the necessary safety integrity, including:

- Management of functional safety and safety lifecycle requirements (Clauses 5 and 6)
- Verification (Clause 7)
- The hazard and risk analyses leading to the specification of safety requirements (Clauses 8 to 10)
- Design, installation, commissioning and validation of the safety-related systems, including the planning requirements for these activities (Clauses 11 to 15)
- Operation and maintenance, modification and retrofit, decommissioning or disposal of safety-related systems (Clause 16 to 18)
- Documentation requirements (Clause 19)

AS(IEC)61511-2 provides a guideline for the application of AS(IEC)61511-1 and AS(IEC)61511-3 provides guidance for the determination of safety integrity levels.

Whilst current guidelines (such as LU07-05 from the New South Wales Department of Primary Industries) do not specifically advocate the use of AS(IEC)61511 in mines, it is clear that there may be situations in mines or in other facilities where its use, in lieu of AS(IEC)61508 / AS(IEC)62061 / ISO13849 / AS4024.1 could be warranted. Examples of such 'process-like' situations could include, but would not be limited to:

- Coal Handling and Processing Plants (CHPP)
- Metallurgical Processing Plants

---

[15] AS(IEC)61511.1 Clause 1.

- Gas Drainage / Flare Systems
- Coal Seam Gas Extraction Networks
- Ventilation Air Methane (VAM) Abatement Systems
- Coal Gasification Plants / 'Char' Plants
- Carbon Sequestration Plants

4.4 ISO13849

ISO13849 Safety of Machinery: Safety-related Parts of Control Systems is a machine-sector specific standard and is intended for use by end-users, machinery designers, control system manufacturers and integrators, and others involved in the design and development of the safety-related parts of machine control systems (SRP/CS). ISO13849 is applicable to electrical / electronic, programmable electronic and "other technology" (ie. hydraulic, pneumatic, mechanical etc…) safety-related systems.

ISO13849 consists of two (2) parts:

- ISO13849-1: Safety of Machinery - Safety-related Parts of Control Systems - Part 1: General Principles for Design
- ISO13849-2: Safety of Machinery - Safety-related Parts of Control Systems - Part 2: Validation

ISO13849-1 was first issued in 1999. It was updated and re-issued in October 2006. ISO13849-2 was first issued in 2003. It was updated and re-issued in October 2012.

In December 2009, ISO13849 (and IEC62061) were adopted as harmonised European Standards under Machinery Directive 2006/42/EC. It was also intended that ISO13849 would replace EN954: Safety of machinery - Safety-related Parts of Control Systems which was withdrawn in November 2009.

The Performance Level (PL) concept is used to describe five (5) classes of safety integrity – PLa, b, c, d and e. PLa describes the lowest level of safety integrity and PLe describes the highest. The PL concept only takes account of 'high' and 'continuous' demand modes of operation of safety-related systems.

For 'high' and 'continuous' demand safety functions, safety integrity is measured by Probability of Dangerous Failure Per Hour (PFH). The maximum PFH's allowable for each level of safety integrity are described as follows in ISO13849-1 Table 3:

PLa: PFH < 0.0001 (or 10^{-4})

PLb: PFH< 0.00001 (or 10^{-5})

PLc: PFH < 0.000003 (or $3x10^{-6}$)

PLd: PFH < 0.000001 (or 10^{-6})

PLe: PFH < 0.0000001 (or 10^{-7})

ISO13849-1 sets out an overall design and development approach and specific requirements to achieve the necessary safety integrity, including:

- Design considerations (Clause 4)
- Specification of safety functions (Clause 5)
- The relationship between categories (CAT) and Performance Level (PL) parameters (Clause 6)

- Fault consideration and fault exclusion (Clause 7)
- Validation (Clause 8)
- Maintenance (Clause 9)
- Documentation (Clause 10)
- Information for use (Clause 11)

Guidelines are also provided for machinery-sector users, including:

- Determination of the required PL (Annex A)
- Safety-related block diagrams (Annex B)
- Calculation of $MTTF_d$ values for components and channels (Annexes C and D)
- Estimation of Diagnostic Coverage (DC) (Annex E)
- Estimation of Common Cause Failure (CCF) (Annex F)
- Avoidance and control of systematic failure (Annex G)
- Examples (Annex H and I)
- Design and development of safety-related software (Annex J)

ISO13849-2 sets out an overall validation approach and specific requirements to achieve the necessary safety integrity, including:

- Validation process (Clause 4)
- Validation by analysis (Clause 5)
- Validation by testing (Clause 6)
- Validation of safety requirements specifications (Clause 7)
- Validation of safety functions (Clause 8)
- Validation of Performance levels (PL's) (Clause 9)
- Validation of environmental requirements (Clause 10)
- Validation of maintenance requirements (Clause 11)
- Validation of technical documentation and information for use (Clause 12)

Guidelines are also provided for machinery-sector users, including:

- Validation tools for mechanical systems (Annex A)
- Validation tools for pneumatic systems (Annex B)
- Validation tools for hydraulic systems (Annex C)
- Validation tools for electrical systems (Annex D)
- Example of validation of fault behavior and diagnostic means (Annex E)

The basic approach of ISO13849 is similar to AS(IEC)61508 in that the emphasis is on risk control relative to the hazards associated with the operation and use of machinery. Following a hazard and risk analysis phase, the safety functions of the machinery control system are identified and their required level of safety integrity is determined via a qualitative risk-based approach.

This results in an allocation of Performance Level (PL) (a, b, c, d or e) for each safety function in a similar fashion to how a SIL would be determined under AS(IEC)61508.

ISO13849 provides guidance on the design and development of the safety-related systems to meet the allocated PL's, in a similar fashion to AS(IEC)61508 / AS(IEC)62061. Importantly, ISO13849 includes requirements for the design and development of safety-related software.

A joint IEC / ISO technical report has been prepared for users of ISO13849 and IEC62061[16]. It explains the relationship between the two standards and the equivalence between PL and SIL. ISO and IEC have both stated that at some point in the future IEC62061 and ISO13849 will be merged into one standard.

Compatibility of ISO13849 and AS4024.1-2006

ISO13849 is no longer compatible with AS4024.1-2006. The content of the 2006 versions of AS4024.1501 and AS4024.1502 were significantly influenced by the content of the ISO13849-1:1999, ISO13849-2:2003, EN954-1:1997 and EN954-2:1999 standards that were available at that time. Importantly, ISO13849-1 was updated and re-issued several months after AS4024.1-2006 was issued. Further updates to ISO13849-1 and ISO13849-2 have occurred since 2006. Furthermore, EN954-1 and EN954-2 were withdrawn in 2009.

Since 2006 ISO13849 has combined the deterministic design approaches of the previous versions of ISO13849 and EN954 with the probabilistic (reliability prediction) design approaches of AS(IEC)61508 and AS(IEC)62061. ISO13849 has in effect become a simplified functional safety standard for machinery applications. The content of AS4024.1501-2006 and AS4024.1502-2006 does reflect the content of the current issues of ISO13849-1 and ISO13849-2.

The prefaces to AS4024.1501-2006 and AS4024.1502-2006 provide the following statement:

"When a new edition of a relevant standard becomes available at the international level, it will be adopted and published within the framework of AS4024 with a minimum delay, so ensuring continued international alignment".

The changes to ISO13849-1 and ISO13849-2 since 2006 are significant and on the basis of the above statement it is therefore expected that they would be adopted by Standards Australia as AS4024.1.1501 and AS4024.1502, or another designation.

It is therefore my view that the approach to the specification and realisation of machinery-sector safety-related systems provided by the latest issues of ISO13849 or AS(IEC)62061 should be used, and the out-dated approaches provided by AS4024.1501-2006 and AS4024.1502-2006 should be avoided where possible.

However, AS4024.1 was in the process of being updated by *Standards Australia* at the time of writing of this book. It is anticipated that AS4024.1 may be re-issued in 2013 or 2014 taking account of the latest issues of ISO12100, ISO14121 and ISO13849.

---

[16] Issued jointly in July 2010 as IEC/TR 62061–1 and ISO/TR 23849. These were produced by a Joint Working Group (JWG) consisting of representatives of ISO/TC199 and IEC/TC44.

4.5 AS4024.1

Note: See Section 4.4 regarding the validity of the design and development approach provided in Part 1501 and 1502 of this standard. It is my view that the approach to the specification and realisation of machinery-sector safety-related systems provided by the latest issues of ISO13849 or AS(IEC)62061 should be used, and the out-dated approaches provided AS4024.1501-2006 and AS4024.1502-2006 should be avoided where possible. AS4024.1 was in the process of being updated by Standards Australia at the time of writing of this book. It is anticipated that AS4024.1 may be re-issued in 2013 or 2014 taking account of the latest issues of ISO12100, ISO14121 and ISO13849.

This Australian standard contains information on the general underlying principles of risk assessment and the achievement of machine safety. It also provides a range of specific information and guidelines on the design of safety-related parts of control systems, human body measurements and ergonomics, machine guarding, control system displays and safety distances.

The Safeguarding Category (CAT) concept is used to describe five (5) classes of safety integrity – CATB, 1, 2, 3 and 4. CATB describes the lowest level of safety integrity and CAT4 describes the highest. The CAT concept does not classify the demand modes of operation of safety-related systems.

Unlike SIL's and PL's, CAT's do not have numerical probability of failure targets. Instead, CAT's are defined by qualitative requirements on system architecture and fault behaviour. The detailed requirements are listed in AS4024.1501 Clause 7, but are summarised as follows:

CATB: The safety-related parts shall be designed, constructed, selected, assembled, and combined in accordance with relevant standards, using 'basic safety principles'[17].

CAT1: The requirements of CATB and, the safety-related parts shall be designed and constructed using 'well-tried components' and 'well-tried safety principles'.

CAT2: The requirements of CATB, the use of 'well-tried safety principles' and, the safety function shall be checked at suitable intervals by the control system.

CAT3: The safety-related parts shall be designed so that a single fault does not lead to loss of the safety function. Where reasonably practicable single faults shall be detected at, or before, the next demand on the safety function.

CAT4: The safety-related parts shall be designed so that a single fault does not lead to loss of the safety function. Single faults shall be detected at or before the next demand on the safety function.

This has been a long-standing standard in Australia, and was updated in 2006 to include 26 parts, which are summarized as follows:

Part 1101: Terminology – Terms and Definitions

Part 1201: General Principles – Basic Terminology and Methodology

Part 1202: General Principles – Technical Principles

Part 1301: Risk Assessment – Principles of Risk Assessment

---

[17] 'Basic safety principles', 'well-tried components' and 'well-tried safety principles' are defined in AS4024.1502 Appendices A to D.

Part 1302: Risk Assessment – Reduction of risks to health and safety from hazardous substances emitted by machinery

Part 1401: Ergonomic Principles – Design Principles

Part 1501: Design of Safety-related parts of Control Systems – General Principles for Design

Part 1502: Design of Safety-related parts of Control Systems – Validation

Part 1601: Design of Controls, Interlocks and Guarding – Guards

Part 1602: Design of Interlocking Devices Associated with Guarding – Principles for Design and Selection

Part 1603: Design of Controls, Interlocks and Guarding – Prevention of Unexpected Start-up

Part 1604: Design of Controls, Interlocks and Guarding – Emergency Stops – Principles for Design

Part 1701: Human Body Measurements – Basic Human Body Measurements for Technological Design

Part 1702: Human Body Measurements – Principles for determining the dimensions required for full body access into openings for whole body access into machinery

Part 1703: Principles for determining the dimensions required for access openings

Part 1704: Human Body Measurements – Anthropomorphic Data

Part 1801: Safety Distances – Safety Distances to Prevent Danger Zones being reached by the Upper Limbs

Part 1802: Safety Distances and Safety Gaps - Safety Distances to Prevent Danger Zones being reached by the Lower Limbs

Part 1803: Safety Distances and Safety Gaps – Minimum Gaps to Prevent Crushing of Parts of the Human Body

Part 1901: Displays, Controls, Actuators and Signals – Ergonomic requirements for the design of displays and control actuators – General principles for human interaction with displays and control actuators

Part 1902: Displays, Controls, Actuators and Signals – Ergonomic requirements for the design of displays and control actuators – Displays

Part 1903: Displays, Controls, Actuators and Signals – Ergonomic requirements for the design of displays and control actuators – Control Actuators

Part 1904: Displays, Controls, Actuators and Signals – Indication, Marking and Actuation – Requirements for Visual, Auditory and Tactile Signals

Part 1905: Displays, Controls, Actuators and Signals – Indication, Marking and Actuation – Requirements for Marking

Part 1906: Displays, Controls, Actuators and Signals – Indication, Marking and Actuation – Requirements for Location and Operation of Actuators

Part 1907: Displays, Controls, Actuators and Signals – System of Auditory and Visual Danger and Information Signals

AS4024.1 includes material drawn from a number of other relevant standards, including:

- AS/NZS4360 (updated in 2009 as AS/NZS/ISO31000:2009) : Risk Management
- EN954:1997: Safety of Machinery - Safety-related Parts of Control Systems - Part 1: General Principles for Design (content adopted by *Standards Australia* in AS4024.1 Part 1501)
- EN954-2:1999: Safety of machinery - Safety-related Parts of Control Systems - Part 2: Validation (content adopted by *Standards Australia* in AS4024.1 Part 1502)
- ISO12100-1:2003: Safety of Machinery - Basic Concepts, General Principles for Design - Part 1: Basic Terminology and Methodology (content adopted by *Standards Australia* in AS4024.1 Part 1201)
- ISO12100-2:2003: Safety of Machinery - Basic Concepts, General Principles for Design - Part 2: Technical principles (content adopted by *Standards Australia* in AS4024.1 Part 1202)
- ISO13849-1:1997: Safety of Machinery - Safety-related Parts of Control Systems - Part 1: General Principles for Design (content adopted by *Standards Australia* in AS4024.1 Part 1501)
- ISO13849-2-2003: Safety of machinery - Safety-related Parts of Control Systems - Part 2: Validation (content adopted by *Standards Australia* in AS4024.1 Part 1502)
- ISO14121:1999: Safety of Machinery - Principles of Risk Assessment (content adopted by *Standards Australia* in AS4024.1 Part 1301)

The basic approach of AS4024.1 is similar to AS(IEC)61508 in that the emphasis is on risk control relative to the hazards associated with the operation and use of machinery. Following a hazard and risk analysis phase, the safety functions of the machinery control system are identified and their required level of safety integrity is determined via a qualitative risk-based approach.

This results in an allocation of CAT B, 1, 2, 3 or 4 for each safety function in a similar fashion to how a SIL would be determined under AS(IEC)61508. However, unlike AS(IEC)61508, AS4024.1 remains prescriptive about how the safety-related parts of the control and protective systems should be designed and integrated, once their necessary level of safety integrity has been determined.

Importantly, where complex or programmable electronic systems are to be used, as in the safety-related parts of machine control systems, these must meet the requirements of AS(IEC)61508 at the requisite Safety Integrity Level (SIL).[18]

AS4024.1 was in the process of being updated by *Standards Australia* at the time of writing of this book. It is anticipated that AS4024.1 may be re-issued in 2013 or 2014 taking account of the latest issues of ISO12100, ISO14121 and ISO13849.

---

[18] See AS4024.1 Part 1202, Clause 5.11.8.

5. An Overview of the Processes

5.1 The AS(IEC)61508 Safety Lifecycle Process

The AS(IEC)61508 standard provides a 16-phase generic "cradle-to-grave" safety lifecycle process, shown here in Figure 5.1 (adapted from AS(IEC)61508.1 Figure 2). This lifecycle provides a rigorous framework of activities, documentation and verifications to ensure that a thorough and repeatable management process is followed. It also ensures that adequate checking and record keeping is performed.

Figure 5.1: AS(IEC)61508 Functional Safety Lifecycle Process

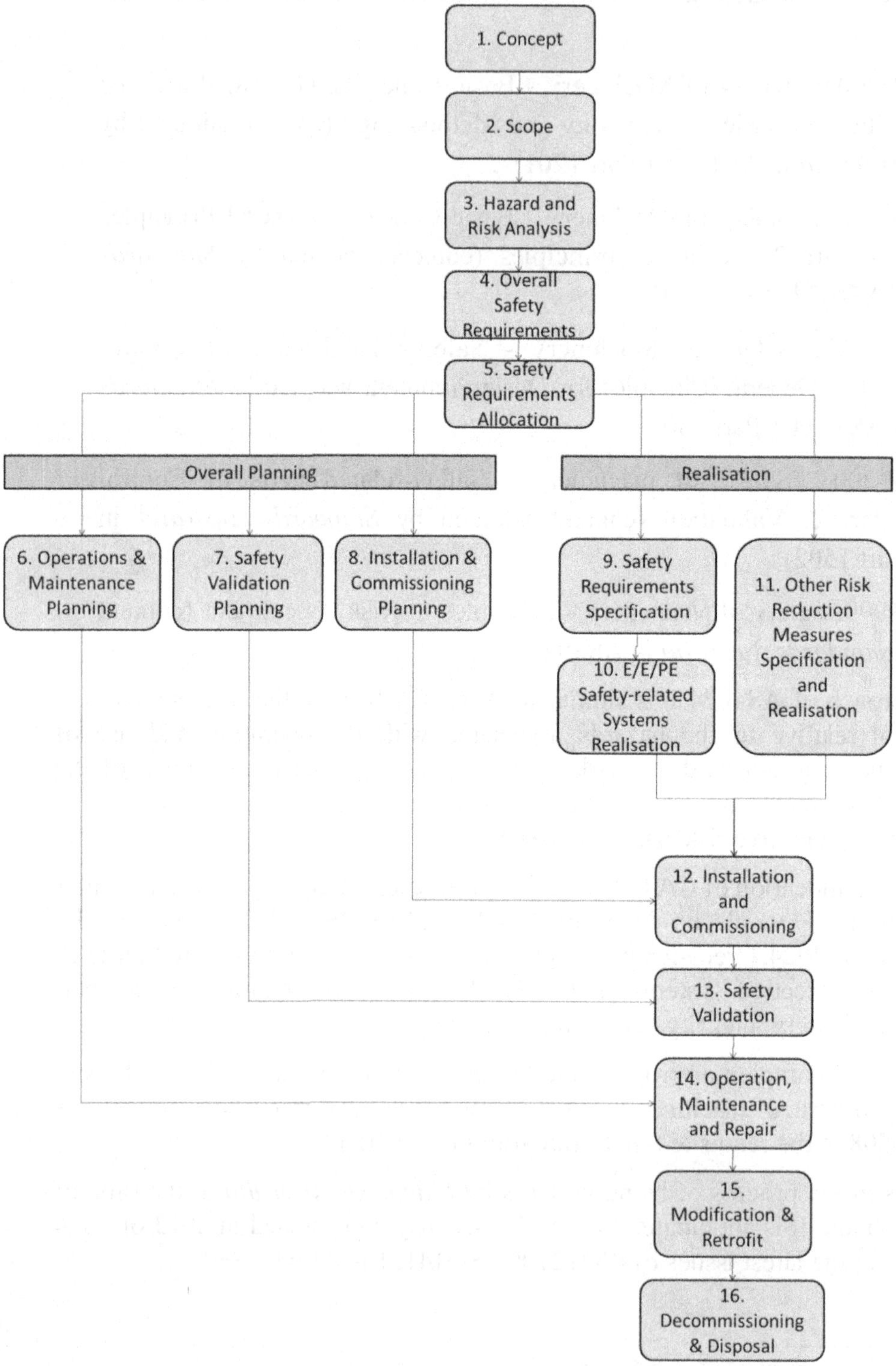

These 16 phases can generally be divided into four fundamental 'stages' – Concept and Scope, Analysis, Realisation, and Operation.

- Concept and Scope: Phases 1 and 2
- Analysis: Phases 3 to 5
- Realisation: Phases 6 to 13
- Operation: Phases 14 to 16

Requirements are provided for all phases except Phase 11 (Other Risk Reduction Measures Specification and Realisation). It is intended that other standards are used for this phase.

5.2 The AS(IEC)62061 Design and Development Process

In AS(IEC)62061, the "realisation" phases of the AS(IEC)61508 lifecycle for E/E/PE safety-related systems (ie. Phases 6, 7, 8, 9, 10 and 12) have been tailored and simplified to meet the specific requirements of machinery-based industries.

AS(IEC)62061 does not provide its own risk assessment process. It calls upon the ISO12100 standard (and by default the ISO14121 standard) for an overall risk assessment process and risk reduction strategy and for determining the need for any safety functions[19]. The risk assessment process assumed by AS(IEC)62061 is shown here in Figure 5.2 (adapted from ISO12100).

Figure 5.2: AS(IEC)62061 Assumed Risk Assessment Process

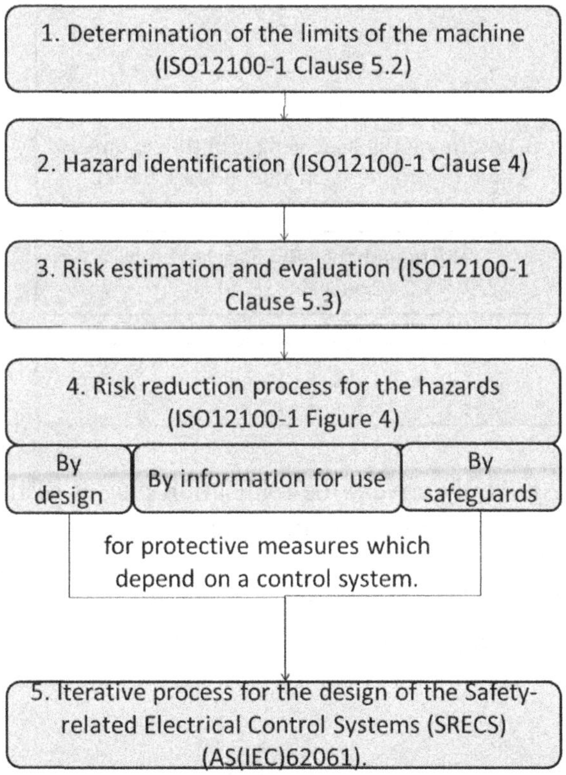

For Step 5, AS(IEC)62061 provides a ten (10) step design and development process for the Safety-related Electrical Control Systems (SRECS). These 10 steps are shown in Figure 5.3 below (adapted from AS(IEC)62061 Figure 2).

[19] In Australia, these standards are currently embodied in the 2006 issue of AS4024.1201, AS4024.1202 and AS4024.1301.

Figure 5.3: AS(IEC)62061 Design and Development Process

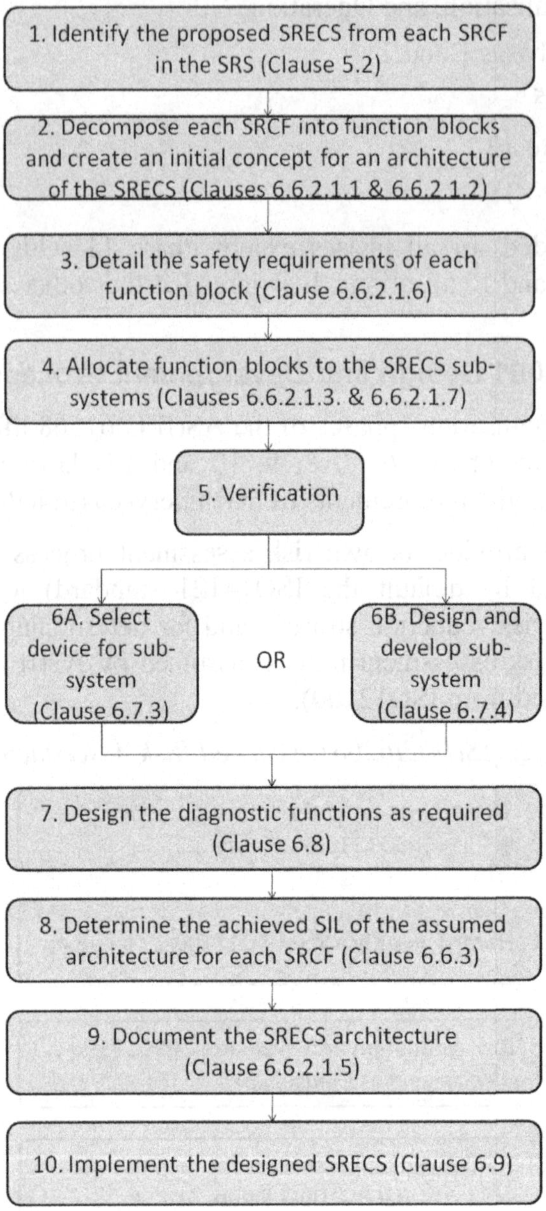

Importantly, AS(IEC)62061 only covers the realisation phases of the safety lifecycle, and only for electrical / electronic / programmable electronic safety-related systems. The use of AS(IEC)62061 alone would not constitute a true 'cradle-to-grave' safety lifecycle process and should therefore be 'sandwiched' within an overall AS(IEC)61508 safety lifecycle process. Section 6.1 provides a basis for doing this.

5.3 The AS(IEC)61511 Safety Lifecycle Process

AS(IEC)61511.1 Figure 8 provides a safety lifecycle process that is very similar to the AS(IEC)61508 process described in Section 5.1. The AS(IEC)61511 safety lifecycle process is shown in Figure 5.4 below.

Figure 5.4: AS(IEC)61511 Safety Lifecycle Process

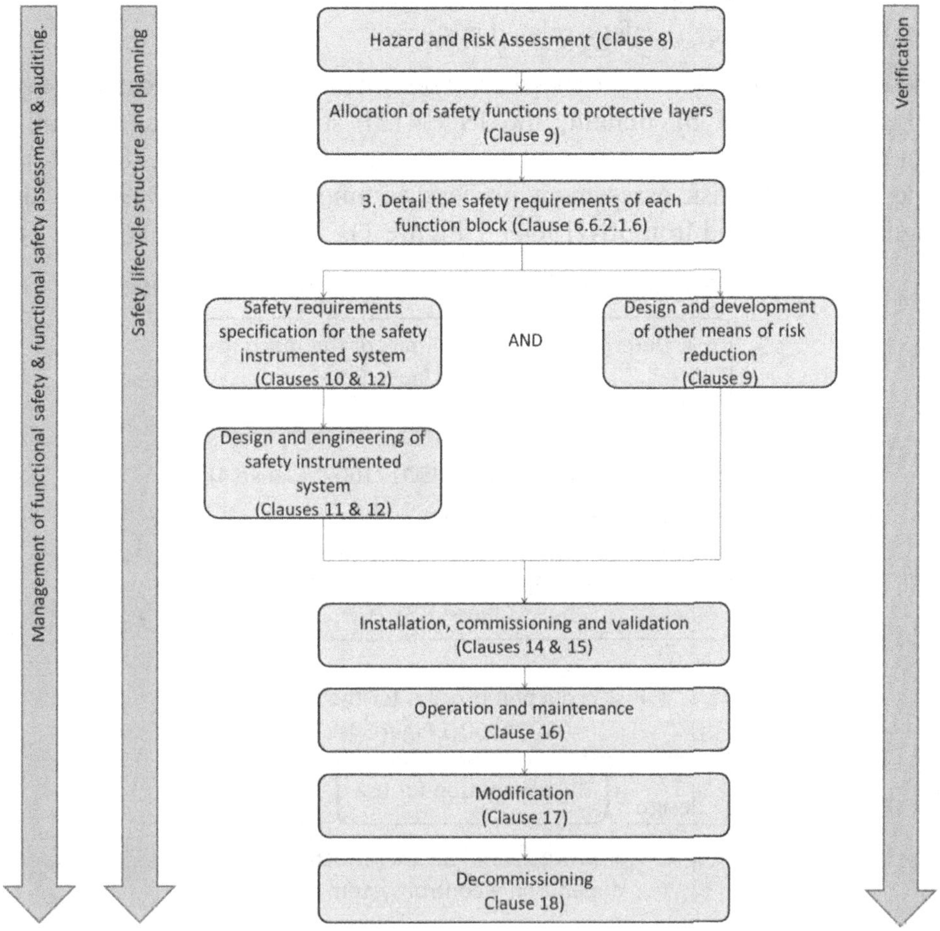

Importantly, AS(IEC)61511 covers all phases of AS(IEC)61508 except Phases 1, 2 and 11, that is:

Phase 1: Concept

Phase 2: Scope

Phase 11: Specification and Realisation of Other Risk Reduction Measures

AS(IEC)61511 may be used as a stand-alone functional safety lifecycle for process applications, however the AS(IEC)61508 standard should be used for those parts of the AS(IEC)61508 safety lifecycle not covered by AS(IEC)61511.

Like AS(IEC)61508, no requirements are provided for Phase 11 (Other Risk Reduction Measures Specification and Realisation). It is intended that other standards are used for this phase.

IEC61511 was in the process of being updated by IEC at the time of writing of this book. It is anticipated that IEC61511 may be re-issued in 2013 or 2014 with a re-issue of AS IEC 61511 thereafter.

5.4 The ISO13849 Design and Development Process

The ISO13849 standard provides a design and development approach that is equivalent to AS(IEC)62061 for the "realisation" phases of the AS(IEC)61508 lifecycle for E/E/PE safety-related systems and "other technology" safety related systems (ie. Phases 6, 7, 8, 9, 10, 11 and 12). Importantly, ISO13849 is applicable to electrical / electronic, programmable electronic and "other technology" (ie. hydraulic, pneumatic, mechanical etc...) safety-related systems.

ISO13849 does not provide its own risk assessment process. It calls upon the ISO12100 standard (and by default, the ISO14121 standard) for an overall risk assessment process, risk reduction strategy and for determining the need for any safety functions[20]. The risk assessment process assumed by ISO13849-1 is shown here in Figure 5.5 (adapted from ISO13849-1 Figure 1).

Figure 5.5: ISO13849 Assumed Risk Assessment Process

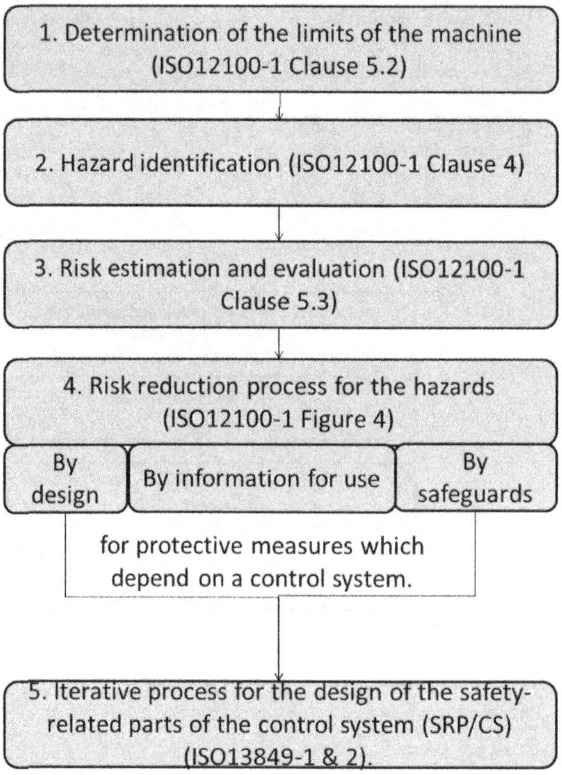

For Step 5, ISO13849 provides a seven (7) step design and development process for the Safety-related Parts of the Control System (SRP/CS). These 7 steps are shown in Figure 5.6 below (adapted from ISO13849-1 Figure 3).

---

[20] In Australia, these standards are currently embodied in the 2006 issue of AS4024.1201, AS4024.1202 and AS4024.1301.

Figure 5.6: ISO13849 Design and Development Process

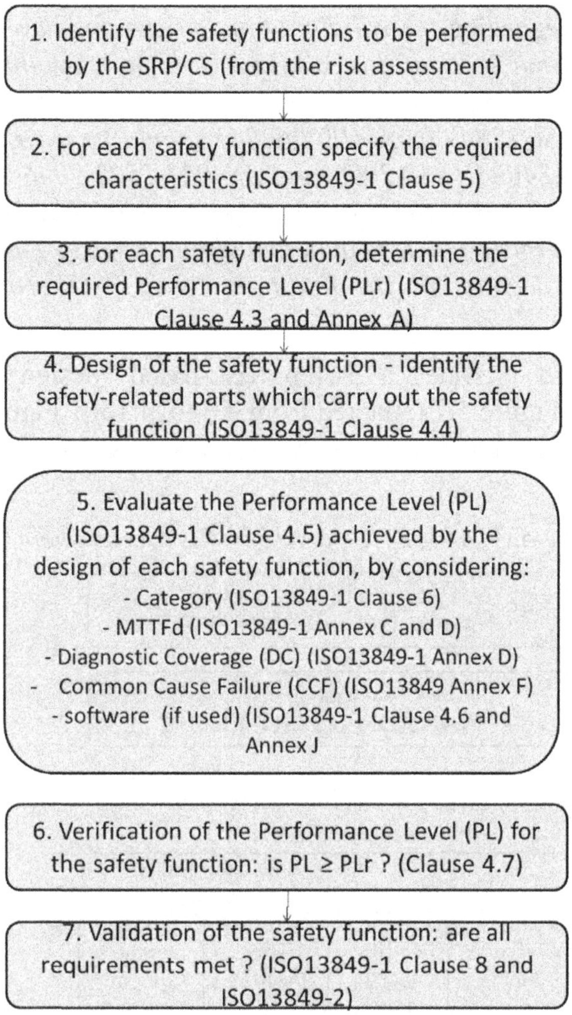

Importantly, ISO13849 only covers the realisation phases of the safety lifecycle . The use of ISO13849 alone would not constitute a true 'cradle-to-grave' safety lifecycle process and should therefore be 'sandwiched' within an overall AS(IEC)61508 safety lifecycle process. Section 6.1 provides a basis for doing this.

5.5 The AS4024.1 Design and Development Process

Note: See Section 4.4 regarding the validity of the design and development approach provided in Part 1501 and 1502 of this standard. It is my view that the approach to the specification and realisation of machinery-sector safety-related systems provided by the latest issues of ISO13849 or AS(IEC)62061 should be used, and the out-dated approaches provided AS4024.1501-2006 and AS4024.1502-2006 should be avoided where possible. AS4024.1 was in the process of being updated by Standards Australia at the time of writing of this book. It is anticipated that AS4024.1 may be re-issued in 2013 or 2014 taking account of the latest issues of ISO12100, ISO14121 and ISO13849.

The AS4024.1 standard provides a 5-step 'risk-based' design and development process, shown here in Figure 5.7 (adapted from AS4024.1501 Figure 1). The specific requirements for safety-related system design and validation are provided in AS4024.1 Part 1502.

Figure 5.7: AS4024.1 Part 1501 Risk Assessment Process

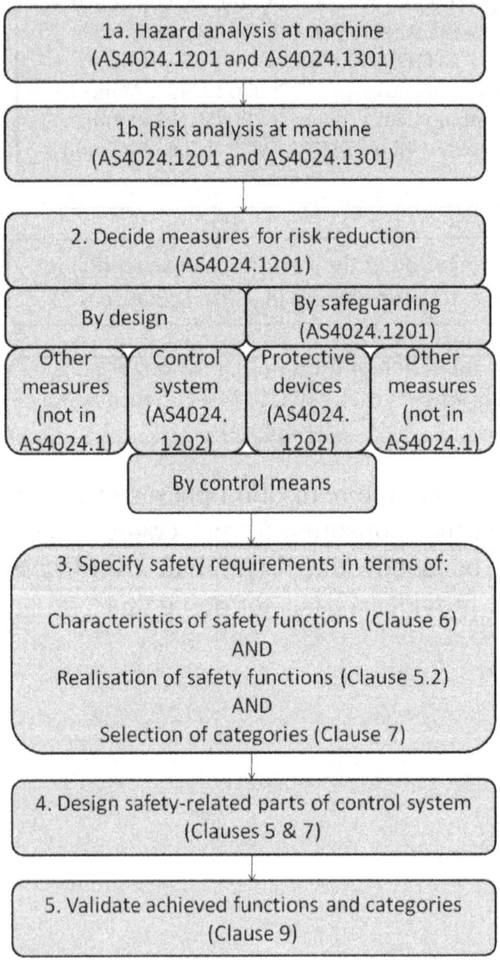

Importantly, AS4024.1 only covers the 'analysis' and 'realisation' stages of the safety lifecycle, and only for electrical / electronic and "other technology" safety-related systems[21]. Nevertheless, the conduct of Steps 1, 2 and 3 are consistent with the

[21] AS4024.1 refers to AS(IEC)61508 for the specific requirements of programmable electronic systems. See AS4024.1 Part 1202, Clause 5.11.8.

general conduct of AS(IEC)61508 Phases 3, 4 and 5 (hazard and risk analysis, safety requirements allocation).

The use of AS4024.1 alone would not constitute a true 'cradle-to-grave' safety lifecycle process and should therefore be 'sandwiched' within an overall AS(IEC)61508 safety lifecycle process. Section 6.1 provides a basis for doing this.

AS4024.1 was in the process of being updated by *Standards Australia* at the time of writing of this book. It is anticipated that AS4024.1 may be re-issued in 2013 or 2014 taking account of the latest issues of ISO12100, ISO14121 and ISO13849.

5.6 The Equivalence of SIL, PL and CAT

SIL, PL and CAT are 'safety integrity target measures'. Each was developed to assist the process of specifying, designing and validating safety functions to the required probability of failure. Whilst SIL, PL and CAT allocations may be derived by a variety of differing methods, and the subsequent design and validation efforts may rely on a variety of differing formal requirements, the fundamental purpose of SIL, PL and CAT is shared.

SIL, PL, CAT and other safety integrity target measures have emerged as the functional safety and machinery standards have evolved and this has been a source of confusion and/or concern to many non-specialists and specialists who use the standards.

A joint IEC / ISO technical report has been prepared for users of ISO13849 and IEC62061[22]. ISO and IEC have both stated that at some point in the future IEC62061 and ISO13849 will be merged into one standard. Many are hopeful that the merging of these two standards will result in a single safety integrity target measure for machinery applications.

The remainder of this section provides a summary of the documented equivalences that exist between SIL, PL and CAT at the time of writing of this book.

The Equivalence of CAT and SIL

The equivalence between the AS4024.1 machinery safeguarding categories (CAT B, 1, 2, 3 and 4) and AS(IEC)61508 / AS(IEC)62061 Safety Integrity Levels (SIL 1, 2 and 3) was drawn in AS(IEC)62061-2006 Table 6, shown below in Table 5.1.

Table 5.1: Equivalence of SIL's and CAT's

Category	Hardware Fault Tolerance (HWFT)	Safe Failure Fraction (SFF)	Maximum SIL Claim Limit according to the architectural constraints
	It is assumed that sub-systems with the stated category have the characteristics given below.		
1	0	SFF < 60%	No SIL (or <SIL1)
2	0	60% ≤ SFF < 90%	SIL1
3	1	SFF < 60%	SIL1
	1	60% ≤ SFF < 90%	SIL2
4	>1	60% ≤ SFF < 90%	SIL3
	1	SFF ≥ 90%	SIL3

[22] Issued jointly in July 2010 as IEC/TR 62061-1 and ISO/TR 23849. These were produced by a Joint Working Group (JWG) consisting of representatives of ISO/TC199 and IEC/TC44.

This table implies that a safety function designed to ISO13849-1:1999 and validated to ISO13849-2:2003[23] may be deemed to be of an equivalent level of safety integrity to one designed to an AS(IEC)61508 / AS(IEC)62061 SIL. The requirements provided in these issues of ISO13849 are embodied in the 2006 issues of AS4024.1501 and AS4024.1502. It would therefore follow that design and validation to AS4024.1501 and AS4024.1502 would be effectively equivalent to the use of ISO13849-1:1999 and ISO13849-2:2003.

However, this equivalence is now outdated due to the updating of ISO13849 that has occurred since 2006 and the replacement of Safeguarding Categories (CAT's) with Performance Levels (PL's) as the safety integrity target measure in that standard[24].

The Equivalence of PL and SIL

The equivalence between the ISO13849 Performance Levels (PL a, b, c, d, and e) and AS(IEC)61508 / AS(IEC)62061 Safety Integrity Levels (SIL 1, 2 and 3) has been drawn in issues of ISO13849-1 (Tables 3 and 4) since 2006, as shown below in Table 5.2.

Table 5.2: Equivalence of PL's and SIL's

Performance Level (PL)	Probability of Dangerous Failure per Hour (PFH)	Equivalent Safety Integrity Level (SIL)
a	$10^{-5} \leq PFH < 10^{-4}$	No SIL (or <SIL1)
b	$3x10^{-6} \leq PFH < 10^{-5}$	SIL 1
c	$10^{-6} \leq PFH < 3x10^{-6}$	SIL 1
d	$10^{-7} \leq PFH < 10^{-6}$	SIL 2
e	$10^{-8} \leq PFH < 10^{-7}$	SIL 3

This table shows that the basis of the equivalence between PL and SIL is Probability of Dangerous Failure per Hour (PFH). PFH is the probability of failure measure used for high demand and continuous mode safety functions. The scope of ISO13849 and AS(IEC)62061 both cover high demand and continuous demand mode safety functions, but not low demand safety functions. Low demand mode safety functions use Probability of Failure on Demand (PFD) as the probability of failure measure. Therefore, there is no specific guidance in post-2006 issues of ISO13849-1 regarding the equivalence of PL's and SIL's for low demand safety functions.

However, I suggest that a common-sense should be applied, as follows:

- It is reasonable to expect that mining and industrial machinery will require a mix of low demand, high demand and continuous mode safety functions
- The PL / SIL equivalences stated in the latest issue of ISO13849-1 should be taken to be valid for high and continuous demand mode safety functions
- The PL / SIL equivalences stated in the latest issue of ISO13849-1 may be taken to be valid for low demand mode safety functions, on the condition that a proof-testing regime is established which would allow the achievement of a PFD in the same SIL range.

[23] See AS(IEC)62061-2006 Clause 6.7.6.4.

[24] See Section 4.4 concerning the validity of AS4024.1 design and development approach in the light of changes to ISO13849.

The Equivalence of CAT and PL

With the advent of PL's in the 2006 issue of ISO13849-1, the use of CAT as a stand-alone safety integrity target measure ceased in that standard. CAT is now one of the four (4) factors to be considered when determining the PL achieved by a safety function. There is no formally documented direct equivalence between CAT and PL.

However, a broad, 'rule-of-thumb' equivalence between CAT and PL can be practically inferred. PL verification is covered in detail in Section 7.13. Figure 7.16 from that section is repeated here (as Figure 5.8).

Figure 5.8: PL Verification

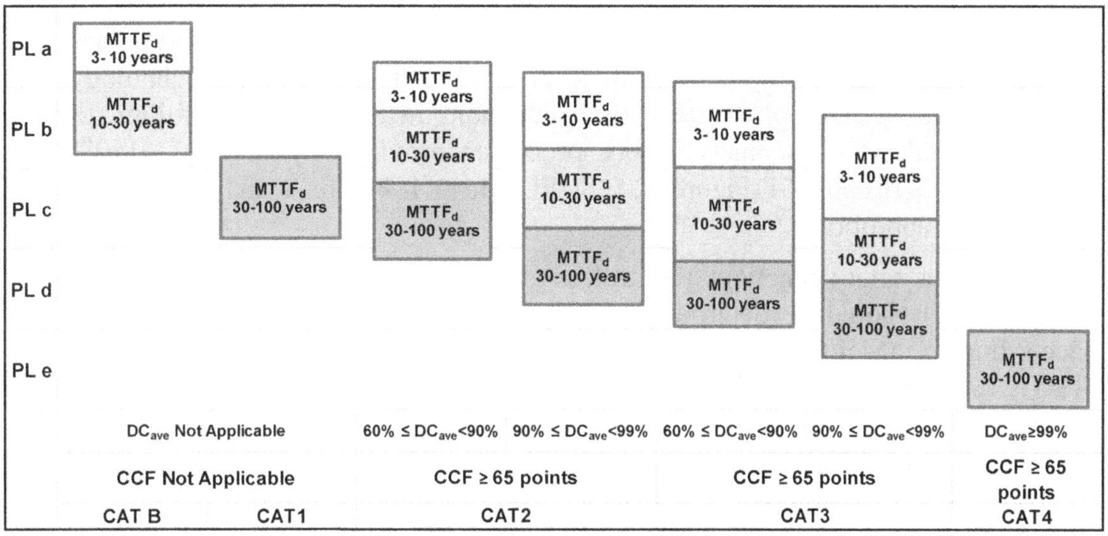

This figure suggests that a safety function with CAT4 architecture can achieve PLe if the Mean Time To Dangerous Failure ($MTTF_d$) is greater than 30 years, Common Cause Failure (CCF) score is greater than 65 and Diagnostic Coverage (DC_{ave}) is at least 99%.

Practical experience has shown that a safety function with CAT4 architecture is also likely to achieve the other three factors required for PLe due to the extent of the redundant design and fault detection inherent in the CAT4 architecture.

If a CAT4 safety function achieves PLe, it would also be equivalent to SIL3, per ISO13849-1 Table 4, thereby maintaining consistency with the outdated relationship between SIL3 and CAT4 provided in ISO13849-1:1999 and AS(IEC)62061-2006 (as shown in Table 5.1).

Similar arguments can be made for CAT3 (equivalence to PLd or SIL2), CAT2 (equivalence to PLb/PLc or SIL1) and CATB/CAT1 (equivalence to PLa). These 'rules of thumb' are based on the values for $MTTF_d$, CCF and DC_{ave} which may be expected due to the extent of redundancy and fault detection inherent in the CAT architectures. These arguments are also consistent with the outdated relationships between SIL and CAT provided in ISO13849-1:1999 and AS(IEC)62061-2006 (as shown in Table 5.1).

6. An Integrated Safety Lifecycle Process - General Requirements

6.1 The Combined Safety Lifecycle Process

A combined safety lifecycle process diagram is provided below in Figure 6.1 which shows the 16 phase AS(IEC)61508 safety lifecycle, with the inclusion of AS(IEC)62061, AS(IEC)61511, ISO13849 and AS4024.1, as appropriate. The diagram also shows the flow of information into and out of each of the lifecycle phases.

The underlying assumptions upon which this approach is based are as follows:

1. The specification, design, installation, validation and operation of the Equipment Under Control (EUC), EUC control system, E/E/PE and "other technology" safety-related systems and "other risk reduction measures" are to follow the functional safety / machinery safety processes provided by AS(IEC)61508.1 Figure 2, AS(IEC)62061 Figure 2, AS(IEC)61511.1 Figure 8 and ISO13849-1 Figure 1, as appropriate.
2. The need for safety functions (regardless of whether they are E/E/PE or "other technology") should be determined via a scoping and risk assessment process according to AS(IEC)61508 Clauses 7.2, 7.3 and 7.4, using AS(IEC)61511.1, ISO12100, ISO14121 and AS4024.1 (Parts 1201, 1202 and 1301), as appropriate. Hazard identification should be carried out via a systematic analysis technique. Risk analysis of the identified hazards shall be carried out using the corporate risk matrix or via a quantitative risk analysis. The risk analysis of each hazard must assess likelihood, foreseeable consequences and what the level of risk is expected to be with no risk controls in place, and with the proposed risk controls in place.
3. The functional and safety integrity requirements of each safety function are to be determined via AS(IEC)61508 Clauses 7.5 and 7.6, AS(IEC)62061 Clause 5, AS(IEC)61511 Clause 9, or ISO13849-1 Clause 5, as appropriate. A SIL or PL allocation should be made on each safety function according to an appropriate methodology given in AS(IEC)61508.5, AS(IEC)62061 Annex A AS(IEC)61511.3 or ISO13849-1 Annex A. The functional and safety integrity requirements should be thoroughly documented and communicated to the designers, via a Safety Requirements Specification (SRS), according to AS(IEC)61508.1 Clause 7.10, AS(IEC)62061 Clause 5, AS(IEC)61511.1 Clause 10 or ISO13849-1 Clause 5, as appropriate.
4. The design and development of the hardware elements of E/E/PE safety-related systems should be guided by the methods outlined in AS(IEC)61508.2, AS(IEC)62061 Figure 2, Clause 6 and Annex B, AS(IEC)61511.2 or ISO13849-1 Clause 4.4 and Figure 3, as appropriate.
5. The design and development of the software elements of E/E/PE safety-related systems should be guided by the methods outlined in AS(IEC)61508.3, AS(IEC)62061 Clauses 10, 11, 12 and Annex C, AS(IEC)61511.2 or ISO13849-1 Figure 6, Clause 4.6 and Annex J, as appropriate.
6. The design and development of "other technology" safety-related systems (ie. mechanical, pneumatic, hydraulic etc…) should be guided by the methods outlined in ISO13849-1 Clause 4.4 and Figure 3.

7. The functional safety process and the achievement of specified functional and safety integrity requirements should be verified and validated at key stages in accordance with the requirements of AS(IEC)61508 Clauses 7.14, 7.18 and 8.2, AS(IEC)62061 Clause 8, AS(IEC)61511.1 Clauses 7 and 15 and ISO13849-1 Clause 4.7 and ISO13849-2, as appropriate. The SIL or PL achieved by each safety function should be confirmed via a reliability analysis, conducted in accordance with AS(IEC)61508.6, AS(IEC)62061 Clauses 6.7.5 to 6.7.8 and Annexes D and F, AS(IEC)61511.1 Clause 11.9, or ISO13849-1 Clauses 4.5 and 6 and Annexes C, D, E and F, as appropriate.
8. The safety lifecycle process should be documented via a Safety Lifecycle Dossier, whose contents are specified in AS(IEC)61508.1 Clause 5 and Annex A, AS(IEC)62061 Clauses 7 and 10, AS(IEC)61511.1 Clause 19 or ISO13849-1 Clauses 10 and 11, as appropriate.
9. Documentation for use by operators and maintainers to assure the continued achievement of a SIL or PL should be made available, where appropriate.
10. Follow-on modifications and retrofitting and the ultimate decommissioning of the EUC, its control system, safety-related systems or other risk controls should be subject to a configuration (engineering change) management process, which considers the impact of changes on the overall safety performance.

Appendix B contains a project functional safety checklist which is based on this framework.

Figure 6.1: Combined Safety Lifecycle Process.

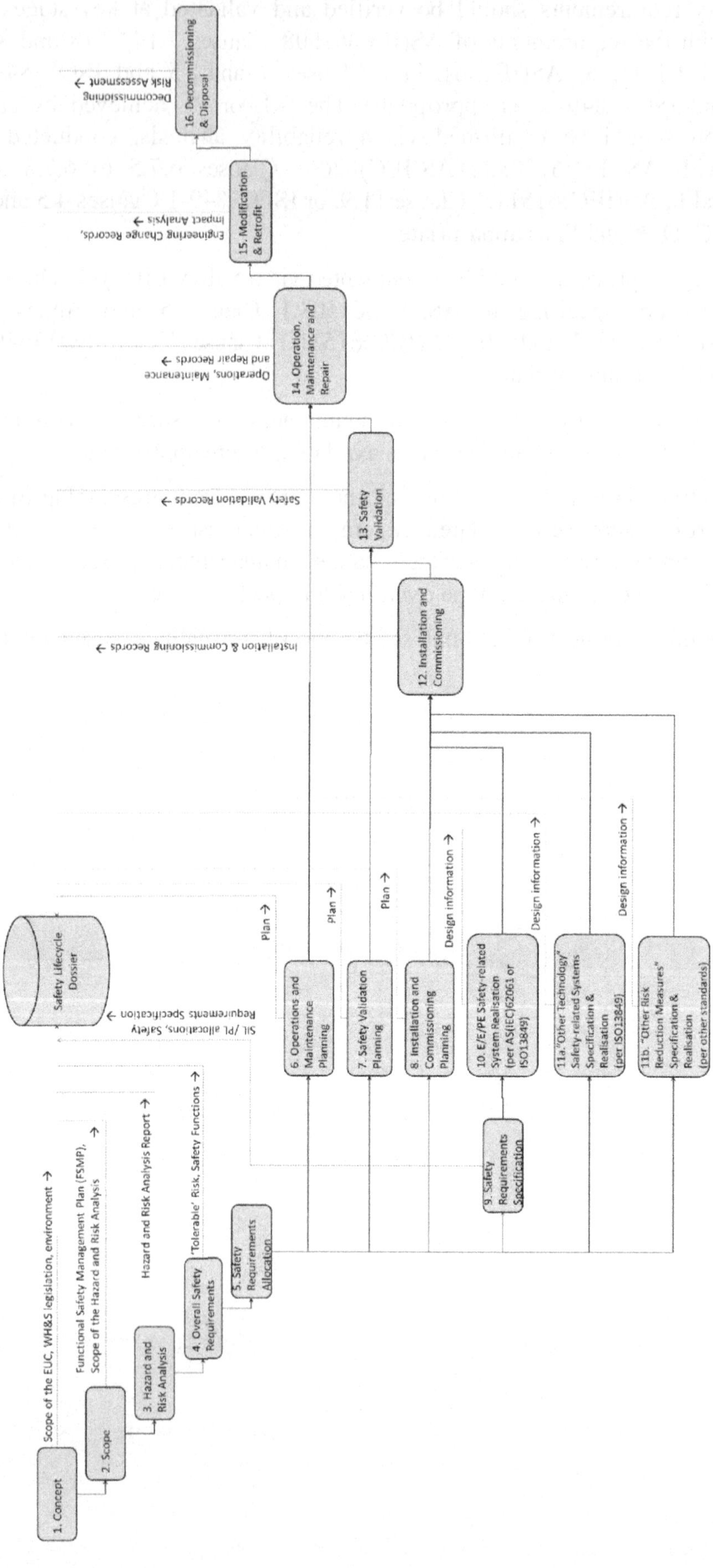

6.2 Overall Responsibilities

The purpose of the safety lifecycle is to establish safety requirements for plant, considering the specific circumstances and risks (eg. environmental, operational, etc...) associated with its use, and then to achieve and maintain those requirements for the duration of the plant's life. Therefore, the following guiding principles should be considered with regard to the division of responsibilities for the safety lifecycle phases and tasks:

- Ownership of the overall safety lifecycle process should belong to, and be driven by, the end-user of the EUC
- The end-user's manager (eg. mine manager), as the end-user's representative, should have overall responsibility for the establishing and maintaining safety lifecycle processes
- Responsibility for certain phases, activities or tasks associated with a particular project or item of plant machine may be assigned to others such as the manager of electrical engineering, manager of mechanical engineering, statutory engineers, project managers or project engineers
- Responsibility for certain phases, activities or tasks may be outsourced to external parties such as design organisations or consultants under certain conditions (see below), but this should occur under the supervision of the end-users manager or their delegate
- Any assignments, delegations or outsourcing that occurs should be documented and communicated to the affected parties.

With regard to the delegation of responsibility for the various safety lifecycle phases, the following should be considered:

- The following phases of the safety lifecycle should be driven by the end-user (eg. the mine) to ensure that the safety requirements that get specified for the EUC are appropriate for the specific application, that is, they are based on a consideration of the hazards associated with the specific environment and operation and the end-user's notion of 'tolerable risk':

  Phase 1: Concept

  Phase 2: Scope

  Phase 3: Hazard and Risk Analysis

  Phase 4: Overall Safety Requirements

  Phase 5: Safety Requirements Allocation

  Phase 9: Safety Requirements Specification

  Clearly, input and effort may be sought from other parties during these phases and this should be done with regard to the guiding principles above. The emphasis is that nothing should occur to that would jeopardise the development of specific safety requirements. For example, the responsibility for performing a SIL/PL allocation for a specific application of plant at a site should not be passed off to the designer. The designer may not have sufficient specific information to do this. The site should perform a risk assessment based on their specific application of the EUC, determine the safety functions and SIL's/PL's on the basis of their notion of 'tolerable risk', and then communicate this to the designer via a Safety Requirements Specification (SRS)

- The establishment of 'tolerable risk' criteria for use in safety requirements allocation (SIL allocation)[25] should be seen as a corporate-level responsibility, in the same way that the establishment of a corporate OH&S risk matrix is seen as a corporate-level function. This approach is consistent with the general requirements of AS/NZS ISO31000:2009 for "setting the context", whereby the criteria for risk evaluation are defined at the start of a risk management process[26]
- The following phases of the safety lifecycle should also be driven by the end-user (eg. the mine) to ensure that the safety requirements are adequately implemented and maintained:

  Phase 6: Operation and Maintenance Planning

  Phase 7: Safety Validation Planning

  Phase 13: Safety Validation

  Phase 14: Operations and Maintenance

  Phase 15: Modification and Retrofit

  Phase 16: Decommissioning

  Again, input and effort may be sought from other parties during these phases according to the guiding principles. The emphasis is that the end-user should drive the activities associated with these phases
- The responsibility for some of the realisation phases of the safety lifecycle may be assigned to other organisations. However it remains the end-user's responsibility to ensure that the other organisation complies with the requirements of those phases:

  Phase 8: Installation and Commissioning Planning

  Phase 10: E/E/PE Safety-related Systems Realisation

  Phase 11: Other Risk Reduction Measures Specification and Realisation

  Phase 12: Installation and Commissioning

6.3 Competencies

Competent people must perform the various activities specified at each step of the lifecycle. It is the end-user's responsibility to ensure that any person assigned to perform activities during a particular step of the safety lifecycle process is competent to do so.

AS(IEC)61508, Part 1, Clauses 6.2.13 to 6.2.15 should be applied to ensure that adequate competencies are applied. Competencies of persons performing work during all steps should be documented. Work to be performed by a sub-contractor or supplier should have a similar level of competency applied, as must the verification, validation and independent assessment activities.

[25] See Chapter 3 regarding the relationship between the 'tolerable risk" approach and the requirements of the harmonised work health and safety legislation, and the additional measures that need to be taken to ensure that the "tolerable risk" approach is consistent with the harmonised laws.

6.4 The Safety Lifecycle Dossier

According to AS(IEC)61508.1 Clause 5 and Annex A, AS(IEC)62061 Clauses 7 and 10, AS(IEC)61511.1 Clause 19 and ISO13849-1 Clause 9 and 10, documentation should be collected and generated during the Safety Lifecycle in order to do the following:

- Manage the functional safety process
- Supply information to those performing tasks during each step in the safety lifecycle
- Verify and validate the level of functional safety achieved
- Provide information for use for operations and maintenance phase
- Allow an independent assessment and judgment of the functional safety process

The repository for safety lifecycle information and documentation is the Safety Lifecycle Dossier. In general, the Safety Lifecycle Dossier must meet the following requirements:

- Documentation generated during the safety lifecycle should be accurate and concise, be easy to understand, suit the purpose for which it is intended and be accessible and maintainable
- A distinction should be made between documentation that is relevant to the end-user and that which is relevant to its design and construction
- Documents should have titles or names indicating the scope of the contents and a revision index to identify different versions of the document

The specific documentation that should be kept at each phase of the functional safety lifecycle is described in Chapter 7 and Appendix B of this book.

For designers of functional safety compliant products such as PLC's, safety relays etc…, AS(IEC)61508.2 Annex D provides detailed requirements for a safety manual of information to enable those products to be integrated into a safety-related system.

Furthermore, AS(IEC)62061 Clause 7 and ISO13849-1 Clause 11 require that information about the installed E/E/PE and "other technology" safety-related systems (ie. "information for use) should be provided by the designer to ensure that the required functional safety is able to be maintained through-life.

In AS(IEC)62061Clause 7, the documentation to be provided by the designer for installation, use and maintenance includes the following. Similar requirements are detailed in ISO13849-1 Clause 11.

- a comprehensive description of the E/E/PE and ""other technology"" safety-related systems equipment, installation and mounting
- a statement of the intended use of the E/E/PE and "other technology" safety-related systems and any measures that can be necessary to prevent reasonably foreseeable misuse
- information on special requirements for use in the physical environment (e.g. lighting, vibration, noise levels, atmospheric contaminants) where appropriate

[26] See AS(IEC)61508.5 Annex C, and particularly Clause C.2.2.

- overview (block) diagrams and circuit diagrams where appropriate
- proof test intervals and expected lifetimes of E/E/PE and "other technology" safety-related systems equipment
- a description (including interconnection diagrams) of the interaction (if any) between the E/E/PE and "other technology" safety-related systems functions and the EUC control system functions
- a description of the measures taken to ensure the separation of the E/E/PE and "other technology" safety-related systems functions from the EUC control system functions
- a description of the safeguarding and of the means provided to maintain safety where it is necessary to suspend operation of the safety functions (e.g. for software changes, repairs, etc.)
- software programming information, where relevant
- description of the maintenance requirements applicable to the E/E/PE and "other technology" safety-related systems including:
    1. a log for recording the maintenance history of the machine
    2. the routine actions which need to be carried out to maintain the functional safety of the E/E/PE and "other technology" safety-related systems, including routine replacement of components with a pre-defined life
    3. the maintenance procedures to be followed when faults occur in the E/E/PE and "other technology" safety-related systems, including - procedures for fault diagnosis and repair, procedures for confirming correct operation subsequent to repairs, maintenance recording requirements
    4. the tools necessary for maintenance and re-commissioning, and the procedures for maintaining the tools and equipment
    5. a specification for periodic testing, preventive maintenance and corrective maintenance

6.5 Verification, Validation and Functional Safety Assessment

AS(IEC)61508 Clauses 7.14, 7.18 and 8.2, AS(IEC)62061 Clause 8, AS(IEC)61511.1 Clauses 7 and 15, ISO13849-1 Clause 4.7 and ISO13849-2 require that Verification and Validation (V&V) and Independent Functional Safety Assessments (IFSA) should be a feature of the safety lifecycle process.

Verification, Validation and Functional Safety Assessment have been defined in AS(IEC)61508.4 as follows:

***validation** - confirmation by examination and provision of objective evidence that the particular requirements for a specific intended use are fulfilled.*

Validation is the activity of demonstrating that the safety-related system under consideration, before or after installation, meets in all respects the safety requirements specification for that safety-related system.

***verification** - confirmation by examination and provision of objective evidence that the requirements have been fulfilled.*

Verification is the activity of demonstrating for each phase of the relevant safety lifecycle, that, for the specific inputs, the deliverables meet in all respects the objectives and requirements set for the specific phase.

functional safety assessment *- investigation, based on evidence, to judge the functional safety achieved by one or more E/E/PE safety-related systems and/or other risk reduction measures.*

Functional Safety Assessment is generally conducted as a high-level audit when an independent judgment is sought about the overall conduct of the safety lifecycle process, its inputs, activities and outputs. Typically, it may be performed at the time of Phase 13 (Safety Validation), prior to commencement of operation of the EUC (and therefore the potential for harm). Further functional safety assessments may also be undertaken at appropriate intervals during Phase 14 (Operations and Maintenance).

The independence of those doing verification, validation and functional safety activities should be commensurate with the minimum recommendations for independence, given in AS(IEC)61508.1 Clauses 8.2.15 to 8.2.18.

For Phases 9, 10 and 11 the independence requirements are set by the SIL allocations for the safety functions. A typical application of these requirements would be:

SIL 1: A person from the same organisation, but not from the group/team that did the work.

SIL 2: A person from the same organisation, but not from the same department / site that did the work.

SIL 3: A person from a separate organization to that which did the work.

For Phases 1 to 8 and 12 to 16 the independence requirements are set by the level of consequence of the hazards in which the safety functions are to be used as risk reduction. A typical application of these requirements would be:

Minor injury: A person from the same organisation, but not from the group/team that did the work.

Multiple serious injuries or a single fatality: A person from the same organisation, but not from the same department / site that did the work.

Multiple fatalities: A person from a separate organization to that which did the work.

The details of those doing verification, validation and functional safety activities, a statement indicating their level of independence, competence and their findings should be documented in the Safety Lifecycle Dossier.

A basic concept for verification, validation and functional safety assessment of the electrical / electronic / programmable electronic safety-related systems, "other technology" safety-related systems and "other risk reduction measures" is as follows. Appendix B contains a Project Safety Lifecycle Checklist. This may be used as the basis for the phase verifications.

Phases 1 – 5 (Concept, Scope, Hazard & Risk Analysis, Safety Requirements)

A verification should be carried when Phases 1-5 have been completed, covering the specific objectives, requirements, inputs and deliverables for those phases.

Phase 9 (Safety Requirements Specification)

A verification should be carried when Phase 9 has been completed, covering the specific objectives, requirements, inputs and deliverables for the phase. For smaller projects, this could be carried out in conjunction with the verification of Phases 1-5.

Phases 6 - 8 (Planning)

A verification should be carried out towards the completion of Phases 6-8 covering the specific objectives, requirements, inputs and deliverables for those phases.

Phases 10 - 11 (Realisation)

A verification should be carried out towards the completion of Phases 10-11 covering the specific objectives, requirements, inputs and deliverables for those phases. This verification may need to be performed in parts (one part for each phase) and it should be scheduled so as to allow time for rectification of the design if deficiencies are found.

In the event that AS(IEC)62061 is used for the design of the electrical / electronic safety-related systems (Phase 10), intermediate verifications are included at Step 5 and Step 8 of the 10-step design and development process (see Section 7.12).

In the event that ISO13849 is used for the design of the electrical / electronic or "other technology" safety-related systems (Phase 10 and 11), an intermediate verification is included at Step 6 of the 7-step design and development process (see Section 7.13).

Phases 12 - 13 (Installation and Commissioning and Safety Validation)

A Safety Validation of the E/E/PE safety-related systems, "other technology" safety-related systems and "other risk reduction measures" (eg. guards, barriers, procedures etc…) should be carried out during, or as part of Phase 12 (Installation and Commissioning).

A verification should also be carried out towards the completion of Phases 12-13 covering the specific objectives, requirements, inputs and deliverables for those phases.

Functional Safety Assessment

A Functional Safety Assessment may be conducted on the project, if an independent judgment is sought as to the overall conduct of the safety lifecycle process. If conducted, it should occur prior to the commencement of Phase 14 (Operations and Maintenance). This should constitute a high-level audit of the application of the safety lifecycle processes, the key activities undertaken, the inputs and outputs, and the contents of the Safety Lifecycle Dossier for Phases 1- 13.

Phase 14 (Operations and Maintenance)

Verifications should be conducted at appropriate intervals during Phases 14 to confirm that the safety lifecycle requirements and safety integrity are being maintained. An appropriate time would be prior to re-registration of the EUC. These should constitute verifications covering the specific objectives, requirements, inputs and deliverables for those phases.

Phase 15 (Modification and Retrofit)

A verification should be carried out towards the completion of any modification and retrofit activity covering the specific objectives, requirements, inputs and deliverables for Phase 15.

Phase 16 (Decommissioning)

A verification should be carried out during decommissioning covering the specific objectives, requirements, inputs and deliverables for Phase 16.

Details on the specific requirements of the verification, validation and functional safety assessment activities should be included in plans in the Safety Lifecycle Dossier.

7. An Integrated Safety Lifecycle Process – Detailed Requirements

7.1 AS(IEC)61508 Phase 1 - Concept

Key Objective/s:

Phase 1 is primarily concerned with gaining an understanding of the Equipment Under Control (EUC) and its environment (ie. physical, legislative etc…) sufficient to enable other safety lifecycle activities to be planned and carried out.

Responsible Entity:

Within the context of a typical mining plant project, the conduct of this phase should be the responsibility of the end-user of the EUC – eg. the mining company.

Activities to be Undertaken:

From AS(IEC)61508.1 Clause 7.2.2, the following should occur to initiate the safety lifecycle activities:

1. A thorough familiarity should be acquired of the proposed EUC, its anticipated control functions and its physical environment.
2. The likely sources of hazard should be determined.
3. Information about hazard sources should be obtained (eg. toxicity, explosive conditions, electrical, corrosiveness, reactivity, flammability etc…).
4. Information about the current safety regulations (national and international) pertaining to the type of EUC (eg. winder, conveyor, mobile plant etc…) should be obtained.
5. Hazards due to interactions with other EUC's (already installed, or to be installed) in the proximity of the EUC should be considered.

Documentation to be Kept:

The information acquired during this phase should be documented and maintained throughout the overall safety lifecycle. The Safety Lifecycle Dossier should be initiated with the following information:

1. Functional specifications, overview drawings and concept schematics of the EUC and its control system, if available.
2. Previous risk assessments conducted by the EUC supplier and/or other users.
3. Literature review of previous safety incident and accident data.
4. MSDS for hazardous materials.
5. Copies of relevant extracts from safety legislation, regulations, codes of practice and safety alerts pertaining to the type of EUC.
6. Proposed layout drawings and concept of operations for the EUC at the site, including its interactions with other plant / EUC's and with people (including operators, workers, maintainers and the public).

7.2 AS(IEC)61508 Phase 2 - Scoping

Objective/s:

AS(IEC)61508 Phase 2 is primarily concerned with determining the functional boundary of the EUC and the EUC control system, and with specifying the scope of the hazard and risk analysis to be conducted during Phase 3.

Responsible Entity:

As with Phase 1, the conduct of this phase should be the responsibility of the end-user of the EUC – eg. the mining company.

Activities to be Undertaken:

From AS(IEC)61508.1 Clause 7.3.2, the following should occur in order to build upon the information gained in Phase 1 and to prepare for Phase 3:

1. The physical environment, including the EUC and EUC control system to be included in the scope of the hazard and risk analysis should be specified.
2. The physical equipment, including the EUC and the EUC control system to be included in the scope of the hazard and risk analysis should be specified.
3. The external events to be included in the hazard and risk analysis should be specified.
4. The sub-systems associated with the hazards should be specified.
5. The type of accident-initiating events that need to be considered (eg. component failures, human error, dependent failures) should be specified.

Furthermore, AS(IEC)62061 specifies that a Functional Safety (Management) Plan (FSMP) should be drawn up early in the safety lifecycle for each project and updated as necessary. Phase 2 is an appropriate point for this plan to be produced as Phase 3 marks the commencement of the risk assessment and functional safety analysis activities. In particular, the FSMP should address the following (per AS(IEC)62061 Clause 4.2):

1. Describe the overall policy and strategy to achieve the functional safety requirements.
2. Identify the relevant safety lifecycle activities that are to be conducted.
3. Identify the persons and organizations responsible for carrying out the safety lifecycle activities,
4. Identify the procedures for recording and maintaining information relevant to functional safety,
5. Describe the overall strategy to achieve functional safety for the application software.
6. Describe a strategy for configuration (engineering change) management.
7. Establish the need for a verification plan.
8. Establish the need for a validation plan.

Documentation to be Kept:

The information acquired should be maintained throughout the overall safety lifecycle. The Safety Lifecycle Dossier should be updated with the following information:

1. Functional Safety Management Plan (FSMP). This plan should also refer to sub-ordinate plans as follows:
2. Verification Plan. This plan may be developed at the same time as the FSMP and could be included within it. The concept for a series of verification activities is provided in Section 6.5. The Verification Plan should include:
a. The scope of the verification activities to be conducted.
b. Details of when the verifications should take place.
c. Details of the persons, departments or units who should carry out the verifications.
d. The selection of verification strategies and techniques.
e. The selection and utilisation of test equipment.
f. Acceptance criteria.
g. The means to be used for the evaluation of verification results.
3. Validation Plan. This plan should be developed during Phase 7 (Safety Validation Planning) - see Section 7.10 for details.
4. Scope of the Hazard and Risk Analysis to be conducted on the EUC.

7.3 AS(IEC)61508 Phase 3 - Hazard and Risk Analysis

Objective/s:

AS(IEC)61508 Phase 3 is primarily concerned with determining the hazards and hazardous events of the EUC and EUC control system, the event sequences leading to those hazardous events, and the risk associated with the hazardous events. To a large extent, Phase 3 may be carried out using the conventional qualitative or quantitative risk assessment procedures and protocols already used by end-users.

Responsible Entity:

The conduct of this phase should be the responsibility of the end-user of the EUC – eg. the mining company. This approach is recommended because the Hazard and Risk Analysis should take account of how the EUC is intended to be used by the particular end-user, including the specific physical environment, operating procedures and the specific exposures of people to hazards. Input should also be garnered from others, such as the EUC and EUC control system designers, other mines, etc…

Activities to be Undertaken:

From AS(IEC)61508.1 Clause 7.4.2 the following should occur:

1. A hazard and risk analysis should be conducted, based on the scope defined in Phase 2.
2. Consideration should be given to the elimination or reduction of hazards[27].
3. There should be consideration of the hazards that may occur under all 'reasonably foreseeable' circumstances (including fault conditions, misuse and malevolent / unauthorized action).
4. The event sequences leading to the hazards should be determined.
5. The likelihood of the hazardous events should be evaluated.
6. The potential consequences of the hazardous events should be determined.
7. The EUC risk (based on the combination of likelihood and consequence) should be evaluated for each determined hazardous event.
8. The hazard and risk analysis should also consider:
a. The contributing components of each hazard,
b. The consequences and likelihood of each hazardous event,
c. The Necessary Risk Reduction for each hazardous event (based upon a notion of "tolerable risk" – discussed in Section 7.6),
d. The measures used to reduce or remove hazards and risks,
e. The assumptions made during the analysis of the risks,

[27] See Chapter 3 regarding the requirements of the harmonised work health and safety legislation, and the additional measures that need to be taken to ensure that this phase is consistent with the harmonised laws.

In addition, I recommend that the following actions are taken to ensure that Phase 3 of the functional safety lifecycle is consistent with the requirements of the Model Workplace Health and Safety Act and Regulations 2011[28]:

1. Risk elimination measures must be used against a hazard, where "reasonably practicable", before risk reduction measures are considered.
2. If risk elimination measures are not "reasonably practicable", the hierarchy of risk controls must be used to guide the selection of the most effective risk reduction measures for each hazard.
3. The selection of risk reduction measures for each hazard must encompass all those that are known, available and suitable at the particular time. All that can be done should be done.
4. If the potential severity of harm from the hazard is significant (ie. serious injury or death) then all known, available and suitable risk reduction measures should be implemented, regardless of cost, or the activity giving rise to the hazard should not be undertaken.
5. Safety requirements allocation against a hazard using a 'tolerable risk' approach should not occur until there is a level of confidence that the selection of risk reduction measures to be used against the hazard would allow the "reasonably practicable" test of the PCBU's primary duty to be satisfied.

Examples of overall processes for Hazard and Risk Analysis are provided in Figures 5.2, 5.5 and 5.7. These are based on the ISO12100 and ISO14121 standards and are consistent with the "risk assessment" (ie. identify, analyse, evaluate) and "risk treatment" parts of the AS/NZS ISO31000 risk management process. Various methodologies are available for the conduct of the hazard identification, risk analysis and risk evaluation steps[29], however the these steps should be conducted using systematic techniques (ie. characterised by the use of order and planning; methodical)[30] and utilise a multi-disciplinary study team. It is preferable that an experienced risk assessment facilitator is involved who has safety lifecycle experience.

Comprehensive checklists have been provided in several standards to initiate the identification of EUC hazards, and these are suitable for initiating this task. For machinery applications, it is suggested that a checklist from one of the following references should be used to initiate the Hazard Analysis.

- ISO14121-1 Annex A and ISO14121-2
- AS4024.1 Part 1301 Annex B, or

[28] See Chapter 3 regarding the requirements of the harmonised work health and safety legislation, and the additional measures that need to be taken to ensure that this phase is consistent with the harmonised laws.

[29] See AS/NZS ISO 31010 Risk management – Risk assessment techniques or HB89-2012 Risk management – Guideline on risk assessment techniques.

[30] Examples of systematic techniques include (but are not limited to) Hazard and Operability Study (HAZOP), Failure Modes and Effects Analysis (FMEA), Fault Tree Analysis (FTA), Event Tree Analysis (ETA), Bowtie analysis, Preliminary Hazard Analysis (PHA) and Structured 'What-if' Techniques (SWIFT).

- MDG1010 Appendix A.4 (a risk assessment guideline used by the mining industry in New South Wales)

Risk Analysis of the identified hazards should be carried out using the end user's standard risk matrix. The Risk Analysis of each hazard should establish its likelihood, the potential exposure of people, and the foreseeable consequences. It is recommended that the health and safety consequences of hazards are assessed on the basis of "Maximum Reasonable Consequence".

Risk controls which are intended to eliminate hazards, reduce the likelihood of their occurrence or mitigate their consequences should be documented as part of the Hazard Risk Analysis. Those risk controls which are comprised of E/E/PE or "other technology" safety-related systems should be earmarked for further analysis in Phase 4 and 5.

The Risk Analysis should establish two evaluations of the risk posed by the hazard:

- Firstly, the inherent level of risk that arises from the hazard (which may be caused by the EUC, the EUC control system, the environment or from human factors) (ie. the EUC risk). EUC risk does not take account of any of the proposed risk controls to be put in place to eliminate or mitigate the hazard
- Secondly, the level of risk that would exist with all of the proposed risk controls in place.

The Hazard and Risk Analysis results may be presented using a variety of formats, depending on the methodologies used. Several examples of a Hazard and Risk Analysis are given in Figures 7.1a, 7.1b and 7.1c below:

Figure 7.1a: Example Hazard and Risk Analysis (narrative presentation style)

Hazard:	Unplanned Start-up of Mobile Plant.
Cause/s:	Mechanical throttle linkages malfunction. Electrical / electronic control system malfunction.
Likelihood (without any risk controls in place): Frequent (up to **1** event per year)	
Consequence:	Unplanned movement of machine. Personnel may be present at the time of the hazard. Potential for severe irreversible disability or fatality. Equipment damage.
Initial Risk:	Severe
Risk Controls:	Design of Traction Control System (including mechanical throttle linkages) Periodic inspection and testing Emergency Stop Procedures for operating in the vicinity of mobile plant (administrative)
Target Likelihood (with all risk controls in place): Remote (less than **0.00001** events per pear)	
Necessary Risk Reduction: 100,000 (ie. **1 ÷ 0.00001**)	
Safety Functions:	Traction Control Start / Stop (including mechanical throttle linkages) Emergency Stop
Residual Risk (with all controls in place): Low	

Figure 7.1b: Example Hazard and Risk Analysis (tabular presentation style)

Study Item / Node Description	Deviation / Guideword	Hazard Description	Cause/s	Description of Potential Consequences	Initial Risk Analysis (no risk controls in place)			Description of Existing / Proposed Risk Controls	Residual Risk Analysis (with risk controls in place)			Safety Function Description Remarks
					C	L	R		C	L	R	
1. Mobile Plant	Unexpected Start-up / Over-run / Over-speed	1. Unplanned start-up of mobile plant.	1. Mechanical throttle linkages malfunction. 2. Electrical / electronic control system malfunction.	Unplanned movement of machine. Personnel may be present at the time of the hazard - potential for severe irreversible disability or fatality.	Fatality	Frequent	Severe	Design of traction control system, including mechanical throttle linkages. Periodic inspection and testing. Emergency Stop. Procedures for operating in the vicinity of mobile plant.	Fatality	Remote	Low	Traction Control - Start / Stop (including mechanical throttle linkages). Emergency Stop.

Figure 7.1c: Example Hazard and Risk Analysis ('Bowtie' presentation style)

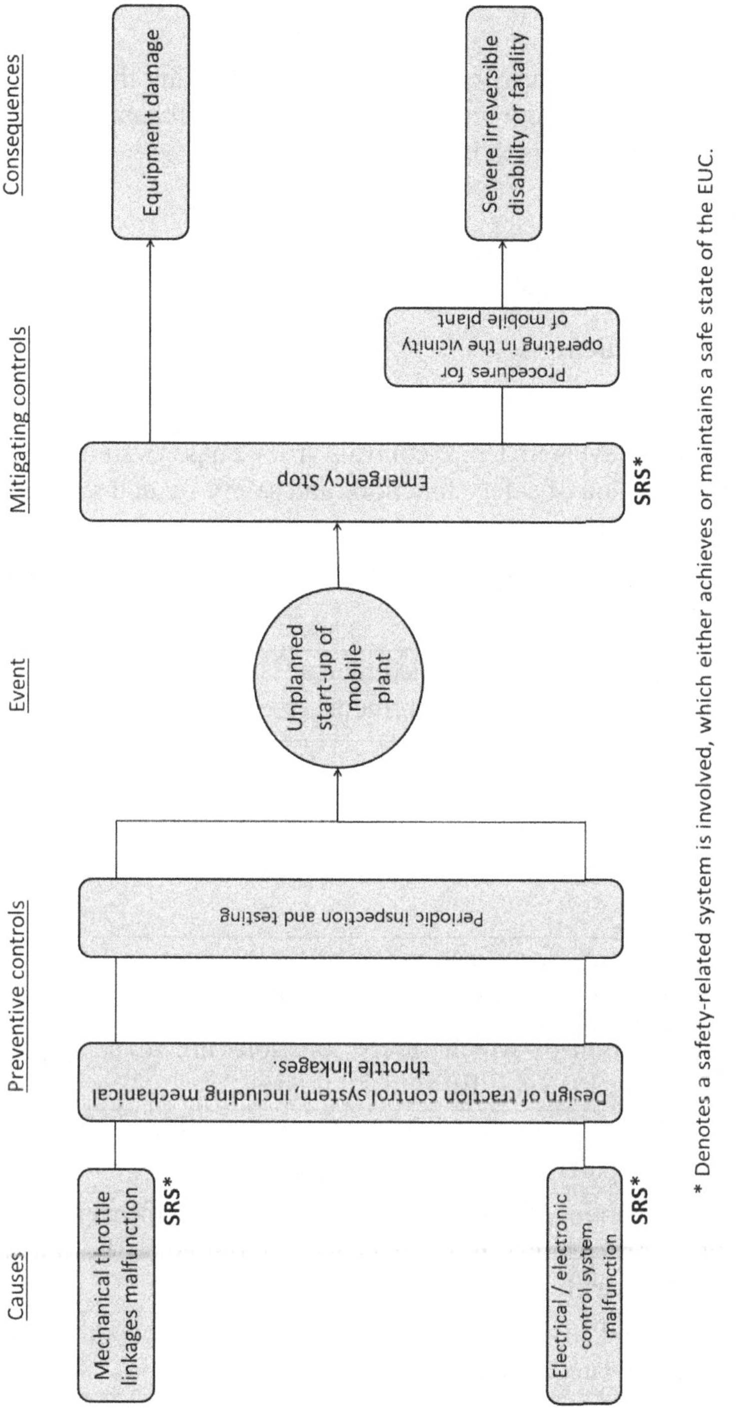

Documentation to be Kept:

The information and results acquired should be maintained throughout the overall safety lifecycle. The Safety Lifecycle Dossier should be updated with the following information:

1.	Hazard and Risk Analysis Report.
2.	Documents used during the hazard and risk analysis (eg. corporate risk matrix, drawings and specifications etc.).

7.4 AS(IEC)61508 Phase 4 - Overall Safety Requirements

Objective/s:

AS(IEC)61508 Phase 4 is primarily concerned with identifying the overall safety function and safety-related system requirements and the establishment of safety criteria or "tolerable risk" targets to be used during Phase 5.

Responsible Entity:

As with Phase 3, the conduct of this phase should be the responsibility of the end-user of the EUC – eg. the mining company.

Activities to be Undertaken:

From AS(IEC)61508.1 Clause 7.5.2 the following should occur:

1. The hazards, risk levels and risk controls from Phase 3 should be used as a basis for identification of safety functions and safety-related systems.
2. A "tolerable risk" frequency should be established so that Safety Requirements Allocation may be performed during AS(IEC)61508 Phase 5 (see Section 7.5) [31].
3. If security threats have been identified then a 'vulnerability analysis' should be undertaken to specify security requirements (see IEC62443).

Documentation to be Kept:

The information and results acquired should be maintained throughout the overall safety lifecycle. The Safety Lifecycle Dossier should be updated with the following information:

1. A Safety Requirements Specification (SRS) for the safety-related systems should be initiated during this phase, and should include, as appropriate:
a. A confirmation of which safety functions are to be implemented (in whole or in part) via E/E/PE technologies.
b. A confirmation of which safety functions are to be implemented (in whole or in part) via "other technologies".
c. A confirmation of the "other risk reduction measures" (ie. guards, barriers, procedures etc…) which are required to be implemented in conjunction with the safety-related systems.
d. The expected environmental conditions and the general requirements on the safety-related systems.
e. A definition of the "safe state" of the EUC.
2. A definition and justification of the 'tolerable risk' target to be used during Phase 5 (see Section 7.6)
3. Vulnerability Analysis Report, if required.

[31] See Chapter 3 regarding the relationship between the "tolerable risk" approach and the requirements of the harmonised work health and safety legislation, and the additional measures that need to be taken to ensure that the "tolerable risk" approach is consistent with the harmonised laws.

7.5 Decision Point - What is Your "Tolerable Risk"?

Integral to the Functional Safety approach is the definition of "tolerable risk" Tolerable risk is defined in order that numerical "safety targets" can be used as a means to compute the required risk reduction to be achieved by safeguards for plant and machinery hazards, and to obtain SIL allocations.

This topic is potentially contentious for industry. Corporations have generally been unwilling to make any statement about what level of risk is "tolerable". Functional Safety is a relatively new concept a number of industries so there is a certain level of misunderstanding about the intention and use of "tolerable risk" targets. The issue is compounded by the recent evolution of the "Zero Harm" philosophy, where the only tolerable risk is one that is zero.[32] Furthermore, it has been observed that AS(IEC)61508 general concept of risk reduction to a notional "tolerable risk" level is at odds with the PCBU's primary obligation (Clause 19) under the Model Act [33].

Nevertheless, non-zero "tolerable risk" targets have already been suggested by various organisations and industry groups throughout the world. The United Kingdom Health & Safety Executive (UK H&SE) and the United Kingdom Institution of Gas Engineers and Managers (IGEM) are two groups who have been at the forefront of these developments.

These two bodies have suggested that the Individual Risk (of death) Per Annum (IRPA) to a worker should be no more than 1 in 1,000 (or 10^{-3}) considering all workplace hazards that they are potentially exposed to. They have also suggested that an IRPA of less than 1 in 1,000,000 (or 10^{-6}) could be considered to be "negligible" or "broadly acceptable"[34].

In Australia, "tolerable risk" targets have already been considered for the mining industry by the Minerals Industry Safety and Health Centre (MISHC):

"Safe" industry should be about 0.005 fatalities / million man-hours, or a factor of ten lower than currently reported in the mining industry.[35]

By 2005 the Australian mining industry Fatal Injury Frequency Rate (FIFR) was approximately 0.05 fatalities per million man-hours. For an individual worker, performing on average 2,000 hours of work per year, this figure translates to an average Individual Risk (of death) Per Annum (IRPA) of about 1 in 10,000 (or 10^{-4}) across the mining workforce. The expected fatality rate is 10 fatalities per 100,000 workers per year.

MISHC suggested that a "safe" industry FIFR target should be 0.005 fatalities per million man-hours, or one tenth of the current FIFR. This figure translates to an

---

[32] Section 8.3 shows how the Functional Safety approach may help to enable the achievement of Zero Harm.

[33] See Chapter 3 regarding the relationship between the 'tolerable risk" approach and the requirements of the harmonised work health and safety legislation, and the additional measures that need to be taken to ensure that the "tolerable risk" approach is consistent with the harmonised laws.

[34] See IGEM SR/15: Integrity of Safety-related Systems in the Gas Industry, 5th Edition (2010), page 5 and Reducing risk, Protecting people (R2P2) – UK HSE's Decision-making Process (2001), pages 44-49.

[35] From page 11, Final Report to Queensland Resources Council (QRC) on underlying Causes of Fatalities and Significant Injuries in the Australian Mining Industry, Minerals Industry Safety and Health Centre (MISHC), March 2005).

average IRPA of 1 in 100,000 (or 10^{-5}) across the Australian mining workforce. The concept of these targets is shown in Figure 7.2.

Figure 7.2: Overall Risk Targets

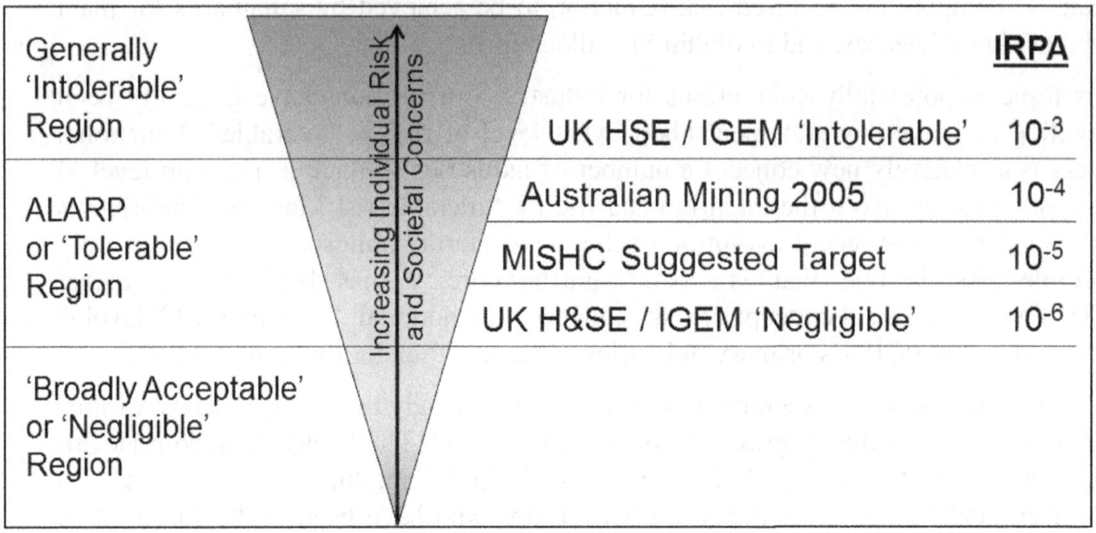

Some gains have been made since 2005, but MISHC's suggested target has not yet been met. Table 3 of Safe Work Australia's report *"Work-related Traumatic Injury Fatalities, Australia 2010-2011"* (December 2012), shown here in Table 7.1 details the fatality rate of Australian mine workers between 2003 and 2011.

Table 7.1: Australian Industry Fatality Rates 2003-2011 (Safe Work Australia 2012)

Industry of employer	2003–04	2004–05	2005–06	2006–07	2007–08	2008–09	2009–10	2010–11
	Fatality rate (deaths per 100 000 workers)							
Agriculture, forestry & fishing	21.54	18.20	16.66	13.41	16.17	19.57	11.38	17.64
Agriculture	*19.90*	*14.01*	*14.64*	*11.10*	*14.29*	*12.28*	*11.08*	*15.65*
Transport, postal & warehousing	13.79	10.04	11.91	14.44	14.35	11.02	8.81	7.20
Road freight transport	*29.69*	*28.71*	*32.92*	*40.20*	*38.11*	*27.83*	*23.38*	*20.50*
Construction	4.91	3.36	4.90	5.31	4.04	4.45	4.08	3.77
Manufacturing	1.65	2.19	2.24	2.93	2.56	2.46	2.39	2.02
Administrative & support services	1.73	2.85	3.38	3.37	3.15	2.62	2.41	2.75
Arts & recreation services	1.95	4.91	1.69	2.76	3.10	1.00	1.01	3.96
Public administration & safety	1.10	2.44	1.80	2.46	0.88	1.94	0.59	1.13
Mining	5.22	7.60	11.64	8.13	6.21	7.18	3.47	3.41
Wholesale trade	3.70	2.63	1.86	1.97	2.81	2.00	1.42	1.45
Retail trade	0.45	0.52	0.77	1.10	0.81	0.41	0.75	0.32
Health care & social assistance*	0.42	0.31	0.29	0.09	0.18	0.26	0.49	0.15
Other services*	1.42	1.69	1.68	1.91	0.65	2.23	0.44	0.44
Professional, scientific & technical services*	0.64	0.61	0.84	0.67	1.02	0.39	0.24	0.23
Accommodation & food services*	0.46	0.59	0.89	0.87	0.43	0.14	0.27	0.26
Information media & telecommunications*	0.46	0.00	0.42	0.40	1.31	0.00	1.40	0.93
Electricity, gas, water & waste services*	7.65	7.35	8.49	3.79	3.54	2.97	2.27	0.66
Rental, hiring & real estate services*	1.69	2.30	3.70	1.50	5.03	2.07	2.16	0.48
Education & training*	0.28	0.14	1.06	0.53	0.50	0.62	0.36	0.12
Financial & insurance services*	0.00	0.00	0.26	0.00	0.00	1.26	0.25	0.00
All industries	**2.82**	**2.58**	**2.87**	**2.90**	**2.75**	**2.64**	**1.99**	**1.93**

The fatality rate has averaged 6.6 fatalities per 100,000 workers over the period, but this occurred within a set of annual results that varied significantly (ie. 3.4 to 11.6 fatalities per 100,000 workers). The average fatality rate for the period 2003-2011 translates to an IRPA of 6.6×10^{-5}. The IRPA's for the individual years ranged between 3.4×10^{-5} and 1.1×10^{-4}.

The MISHC "safe" industry target therefore still represents a substantial improvement on current safety performance (ie. 6.6 times better) in the mining industry and recent gains indicate that it is an achievable target in the medium-term.

The UK H&SE and IGEM suggest that a "broadly acceptable" risk level requires an IRPA of less than 1 in 1,000,000 (or 10^{-6}). This figure translates to a FIFR of less than 0.0005 fatalities per million man-hours, which is one tenth of the above MISHC "safe" industry FIFR target, and one hundredth of the 2005 FIFR performance.

These risk targets consider the overall risk to a worker considering all hazards. However in practice, risks are identified and assessed on an individual hazard basis and only consider those workers who are potentially exposed. For the purposes of the Functional Safety approach these IRPA targets need to be translated to a 'per hazard' basis. A scheme for doing this is provided in Table 7.2. The bolded figures are explained over-page.

Table 7.2: "Per Hazard" Safety Targets

IRPA Target (fatalities / employee / year)	No. of Employees at Site	Site-wide Fatality Target (fatalities / year)	No. of Fatal Hazards Present	Per Hazard Frequency Target (events / year)
10^{-5}	1	10^{-5}	10	10^{-6}
			100	10^{-7}
			1000	10^{-8}
	10	10^{-4}	10	10^{-5}
			100	10^{-6}
			1000	10^{-7}
	100	$\mathbf{10^{-3}}$	10	10^{-4}
			100	$\mathbf{10^{-5}}$
			1000	$\mathbf{10^{-6}}$
	1000	$\mathbf{10^{-2}}$	10	10^{-3}
			100	$\mathbf{10^{-4}}$
			1000	$\mathbf{10^{-5}}$
	10000	10^{-1}	10	10^{-2}
			100	10^{-3}
			1000	10^{-4}

The 'per hazard' safety target takes account of the chosen IRPA target, the size of the workforce and the number of hazards posed to the employees. For a typical mine site with 100 to 1000 employees continuously present (bolded), and with 100 to1000 fatal hazards (bolded) posed to each employee the suggested hazard frequency target is 10^{-4} to 10^{-6} per year (bolded). A nominal target of 10^{-5} per year is suggested.

This is further supported by mining industry guidance provided in MDG1010. An extrapolation of the safety risk matrix suggested in MDG1010, Figure A.9.2 (shown here in Figure 7.3) indicates that for a single fatality consequence, a "Low" risk would only be achieved with a hazardous event frequency less than once per 100,000 years, (or, 10^{-5} / yr).

Figure 7.3: MDG1010 Safety Risk Matrix

	1 x Medically Treatable Injury (MTI)	1 x Compensible Injury (CI) or 10 x MTI's	10 x CI's	1 x Permanent Disablement (PD)	1 x Fatality	10 x Fatalities
Frequent ≥ 1 per year	Medium	High	Very High	Severe	Severe	Severe
Possible: < 1 per year (but ≥ 0.1 per year)	Medium / Low	Medium	High	Very High	Severe	Severe
Unlikely: < 0.1 per year (but ≥ 0.01 per year)	Low	Medium / Low	Medium	High	Very High	Severe
Very Unlikely: < 0.01 per year (but ≥ 0.001 per year)		Low	Medium / Low	Medium	High	Very High
Barely Credible: < 0.001 per year (but ≥ 0.0001 per year)			Low	Medium / Low	Medium	High
< 10^{-4} per year	→	→	→	Low	Medium / Low	Medium
< 10^{-5} per year	→	→	→	→	Low	Medium / Low
< 10^{-6} per year	→	→	→	→	→	Low

This target may be adjusted for those hazards which could result in multiple fatalities, or in less serious safety consequences. For example, a hazard with the potential for ten (10) fatalities could be assigned the target of 10^{-6} per year, one-tenth of the single fatality target. A hazard with the potential for serious injury might be assigned a target somewhere in the range of 10^{-4} to 10^{-5} per year.

7.6 AS(IEC)61508 Phase 5 - Safety Requirements Allocation

Objective/s:

AS(IEC)61508 Phase 5 is primarily concerned with allocating a level of safety integrity to each safety function that is to be carried out by an E/E/PE safety-related system. Phase 5 should also include consideration of the "other technology" (ie. mechanical, pneumatic, hydraulic etc.) safety-related systems and "other risk reduction measures" (ie. guards, barriers, procedures etc.) in order to achieve the required functional safety.

If AS(IEC)62061, AS(IEC)61511, ISO13849-1 or AS4024.1 are to be used within the overall AS(IEC)61508 safety lifecycle, there is a transition from AS(IEC)61508 requirements to the requirements of these standards during this phase.

Responsible Entity:

As with Phases 3 and 4, the conduct of this phase should be the responsibility of the end-user of the EUC – eg. the mining company.

Activities to be Undertaken:

From AS(IEC)61508.1 Clause 7.6.2, AS(IEC)62061 Clause 5, AS(IEC)61511 Clause 9 and ISO13849-1 Clause 4.3 the following should occur:

1. The hazards, risk levels and risk controls from Phase 3 should be used as an input to this step.
2. The concept of "tolerable risk" from Phase 4 should be used as an input to this step (AS(IEC)61508 only) [36].
3. The safety integrity requirements for each safety function (whether E/E/PE or "other technology" based) should be determined by a SIL or PL allocation conducted according to an appropriate method[37].
4. Where a safety function is to be implemented by E/E/PE technologies, then its functional and safety integrity requirements should be specified per AS(IEC)61508.1 Clause 7.10, AS(IEC)62061 Clause 5.2, AS(IEC)61511 Clause 10 or ISO13849 Clause 5, as appropriate.
5. Where a safety function is to be implemented by "other technology" (eg. pneumatic, hydraulic, mechanical etc...) then its functional and safety integrity requirements should be specified per ISO13849-1 Clause 5.
6. If a safety function is intended to be implemented by a combination of "other technology" and E/E/PE parts, then the functional and safety integrity requirements of the individual parts should be specified via a standard most appropriate to the technology. Table 1 of AS(IEC)62061 and Table 1 of ISO13849-1 provide guidance on the applicability of AS(IEC)62061 and ISO13849-1 to E/E/PE and "other technology" parts.

[36] See Chapter 3 regarding the relationship between the "tolerable risk" approach and the requirements of the harmonised work health and safety legislation, and the additional measures that need to be taken to ensure that the "tolerable risk" approach is consistent with the harmonised laws.

[37] See Section 7.7.

7.	Where the proposed set of risk controls for a hazard is insufficient to achieve a "tolerable risk", an additional safety function must be considered.
8.	The functional and safety integrity requirements of each safety function comprising the E/E/PE and "other technology" safety-related systems should be documented in a Safety Requirements Specification (SRS), which should form part of the overall specification of the EUC.
9.	Any "other risk reduction measures" (ie. guards, barriers, procedures etc.) that are intended to work in co-ordination with the safety functions should be defined in the SRS.

In addition, I recommend that the following action is taken to ensure that Phase 5 of the functional safety lifecycle is consistent with the requirements of the Model Workplace Health and Safety Act and Regulations 2011:

1. Safety requirements allocation against a hazard using a 'tolerable risk' approach should not occur until there is a level of confidence that the selection of risk reduction measures to be used against the hazard would allow the "reasonably practicable" test of the PCBU's primary duty to be satisfied.
2. The allocation of levels of safety integrity (ie. SIL, PL or CAT) to any safety-related systems must also take account of the level of safety integrity that would be "reasonably practicable" for a designer to achieve.
3. Where a regulation, code of practice or standard specifies a level of safety integrity or other specific requirements for a particular type of safety-related system, and this is more onerous than the SIL, PL or CAT allocated, then this requirement should take precedence.

Documentation to be Kept:

The information and results acquired should be maintained throughout the overall safety lifecycle. The Safety Lifecycle Dossier should be updated with the following information:

1.	Safety Requirements Allocation Report (if Hazard and Risk Analysis and Safety Requirements Allocation are co-conducted, then a combined report may be provided).

7.7 Decision Point - Choose a Method of SIL or PL Allocation

Mining and industrial plant and machinery can be complex, and are increasingly so. As a general rule, complex systems demand a more rigorous and deeper level of analysis than simple systems. SIL and PL allocation for mining plant and other complex machinery should be no exception to this rule.

The SIL allocation process can be explained by referring to the general concept for risk reduction described in AS(IEC)61508.5 Figure A.1 (shown here in Figure 7.4).

Figure 7.4: AS(IEC)61508 General Concept for Risk Reduction

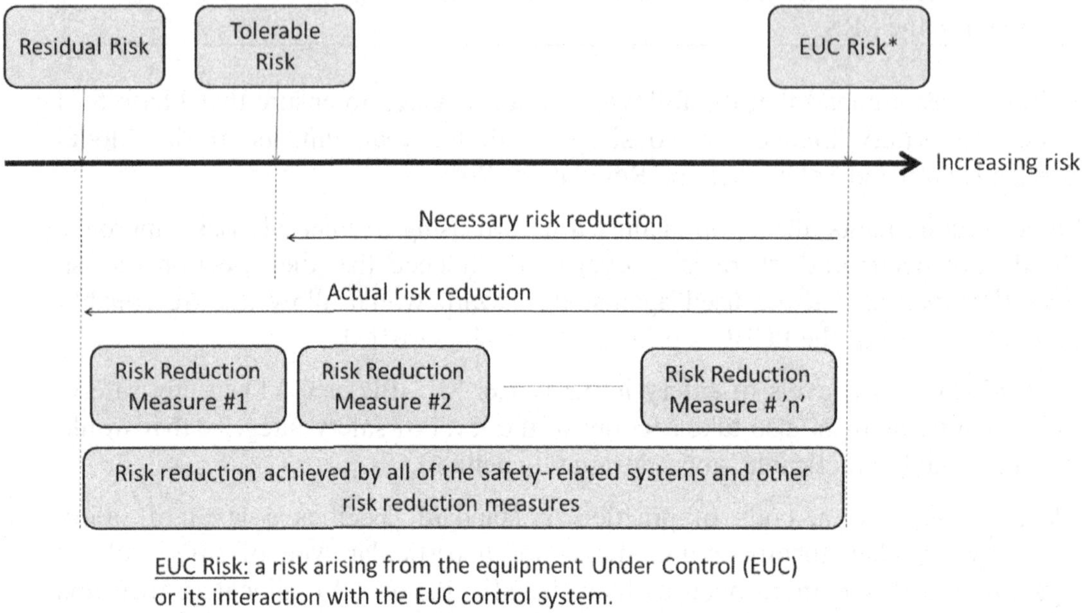

Phases 3 to 5 of the AS(IEC)61508 functional safety lifecycle follow a "tolerable risk" approach to risk management that is consistent with the model of risk management advocated in the AS/NZS ISO31000 Risk Management standard[38].

For each EUC risk that is identified, the level of risk is estimated and then one or more risk reduction measures are selected. In AS(IEC)61508 risk reduction measures are broadly divided into two categories: safety-related systems and "other risk reduction measures". Each risk reduction measure provides its own degree of risk reduction against the EUC risk. Collectively, all of the risk reduction measures provide an overall degree of risk reduction against the EUC risk, known as the "actual risk reduction". As the EUC risk has not been eliminated, a level of "residual risk" remains in place, following the implementation of all of the risk reduction measures.

The aim of this approach to risk management is to apply sufficient risk reduction measures against the EUC risk such that the "actual risk reduction" exceeds the minimum level of risk reduction that would be needed (ie. the "necessary risk reduction") to achieve a notional level of "tolerable risk" which is set by the end-user of the EUC. The notion of "tolerable risk" is discussed in detail in Section 7.5.

[38] See Chapter 3 regarding the relationship between the "tolerable risk" approach and the requirements of the harmonised work health and safety legislation, and the additional measures that need to be taken to ensure that the "tolerable risk" approach is consistent with the harmonised laws.

AS(IEC)61508, AS(IEC)62061, AS(IEC)61511 and other functional safety standards provide a range of methods for SIL allocation, ranging from simple qualitative methods to more complex quantitative methods.

The relevant guidelines for quantitative and semi-quantitative SIL allocation are:

- AS(IEC)61508.5 Annex D (Quantitative Method)
- AS(IEC)61508.5 Annex F (Layer of Protection Analysis (LOPA))
- AS(IEC)61511.3 Annex B (Semi-quantitative Method)
- AS(IEC)61511.3 Annex F (Layer of Protection Analysis (LOPA))

The relevant guidelines for qualitative SIL allocation methods are:

- AS(IEC)61508.5 Annex E (4-parameter Risk Graph)
- AS(IEC)61508.5 Annex G (3-parameter Hazard Event Severity Matrix)
- AS(IEC)62061 Annex A (4-parameter Risk Estimate)
- AS(IEC)61511.3 Annex D (4-parameter Risk Graph)
- AS(IEC)61511.3 Annex C (3-parameter Safety Layer Matrix).

Furthermore, ISO13849-1 Annex A provides a qualitative method for PL allocation which is based on a 3-parameter risk graph. ISO13849-1 does not provide any quantitative or semi-quantitative methods for PL allocation. Should end-users prefer that their suppliers / designers use the ISO13849-1 approach they might consider allocating a SIL to each safety function via a quantitative or semi-quantitative means first, then converting it to the equivalent PL, per Section 5.6 and Table 5.2.

It should be noted that it is sometimes the case that these methodologies may deliver different SIL or PL allocations when applied to the same hazardous scenario. This happens because:

1. Some methodologies involve a 'qualitative' assessment of risk (which is a coarser, or less accurate assessment of risk) and others involve a 'quantitative' assessment of risk (which is a finer, or more accurate assessment of risk).
2. Each methodology has its own evaluation parameters, definitions and parameter selection criteria.
3. Some allocation methods are calibrated to a particular 'tolerable risk' frequency (such as the AS(IEC)61508.5 risk graph), whilst others are not (such as the ISO13849 risk graph).

If this circumstance arises I suggest that an allocation is chosen via the following order of precedence of the methods:

1. "Quantitative" or "semi-quantitative" method that is calibrated to a "tolerable risk" frequency.
2. "Qualitative" method that is calibrated to a "tolerable risk" frequency.
3. "Quantitative" or "Qualitative" method that is not calibrated to a "tolerable risk" frequency.

The final allocation given to the safety-related system must also take account of the level of safety integrity that would be "reasonably practicable" for a designer to achieve.

Examples of safety requirements allocation using the various methodologies are given in Figures 7.5a to 7.5e below.

Figure 7.5a: Quantitative SIL Allocation (using Fault Tree Analysis (FTA))

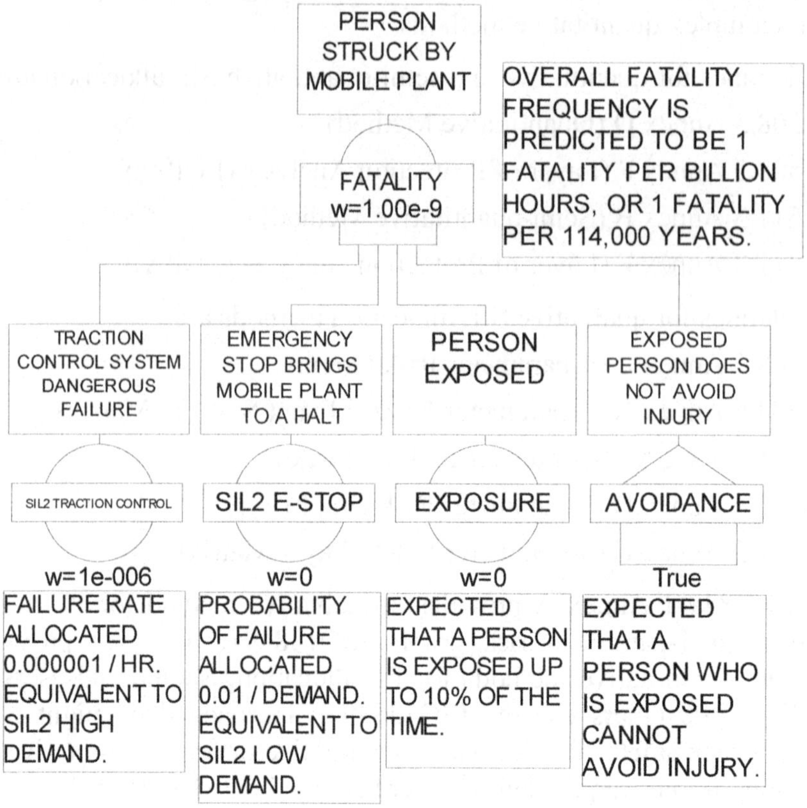

Figure 7.5b: Quantitative SIL Allocation (using Event Tree Analysis (ETA))

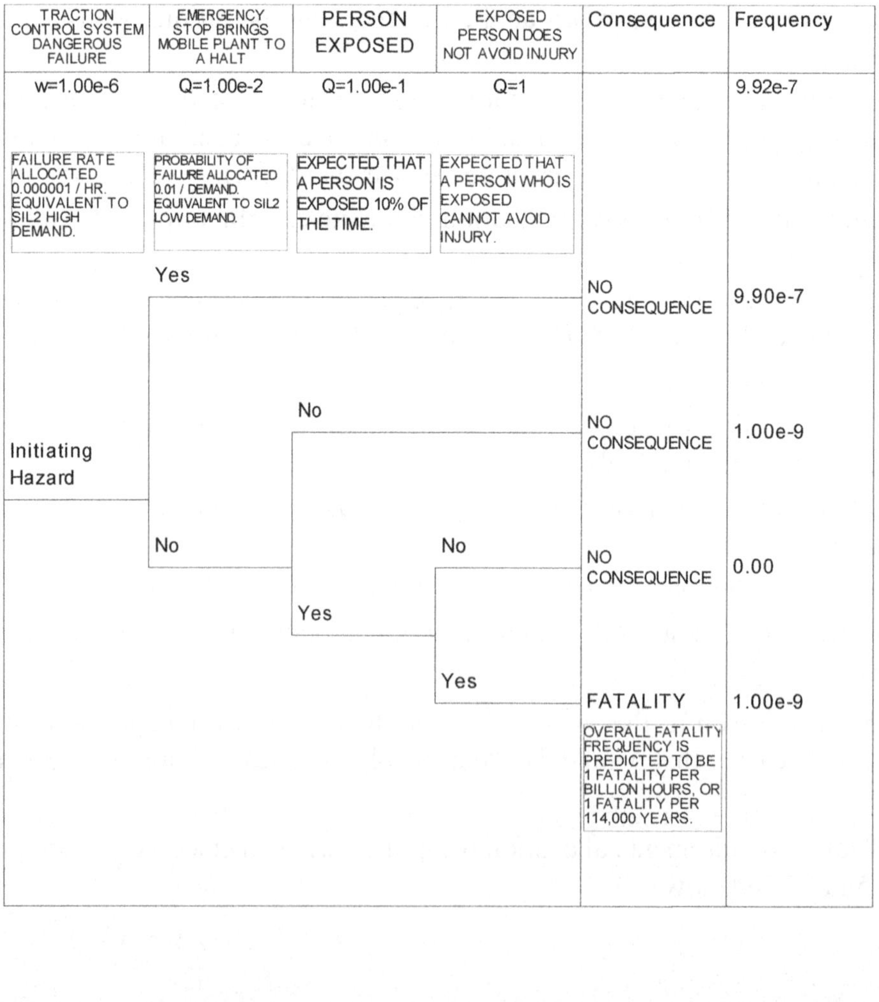

Figure 7.5c: Semi-quantitative SIL Allocation (using (LOPA)

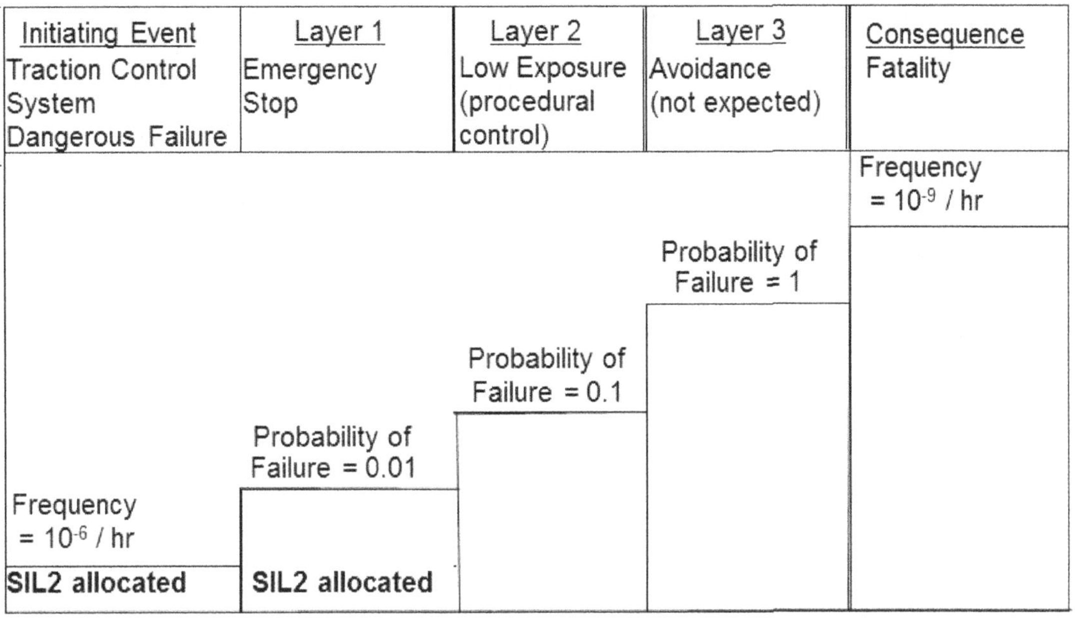

Figure 7.5d: Qualitative SIL Allocation (using 4-parameter Risk Graph)

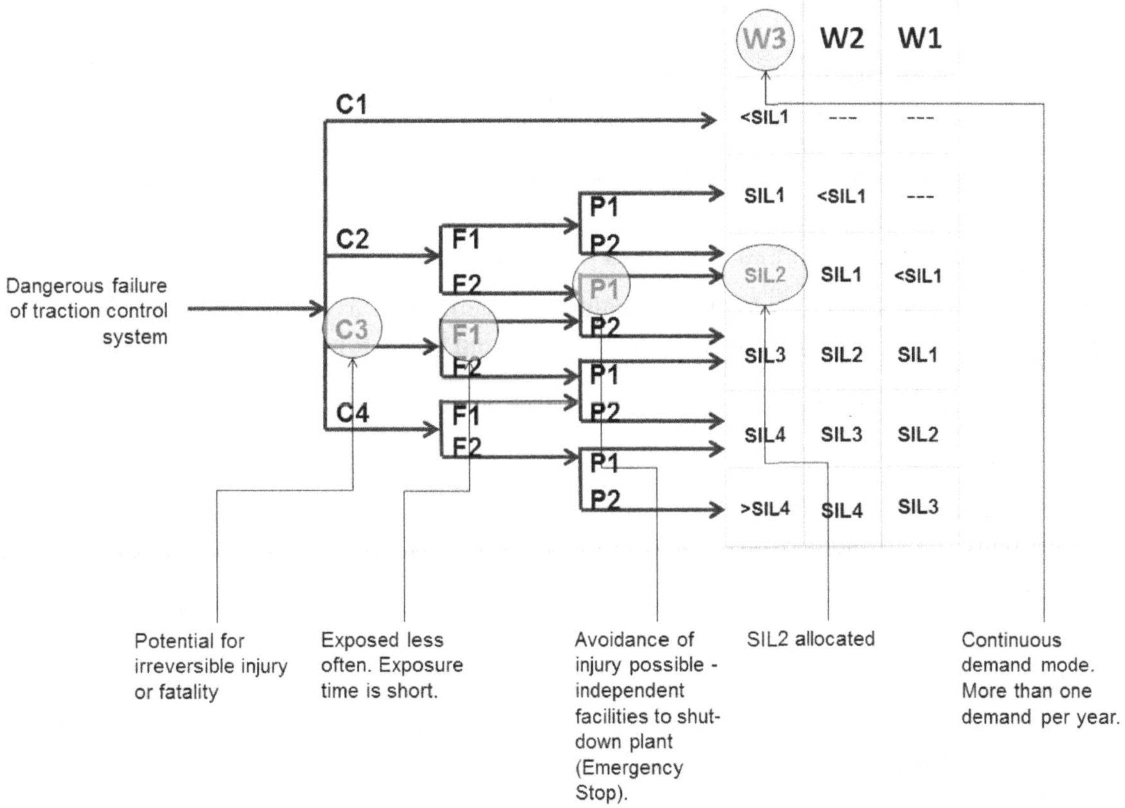

Figure 7.5e: Qualitative PL Allocation (using 3-parameter Risk Graph)

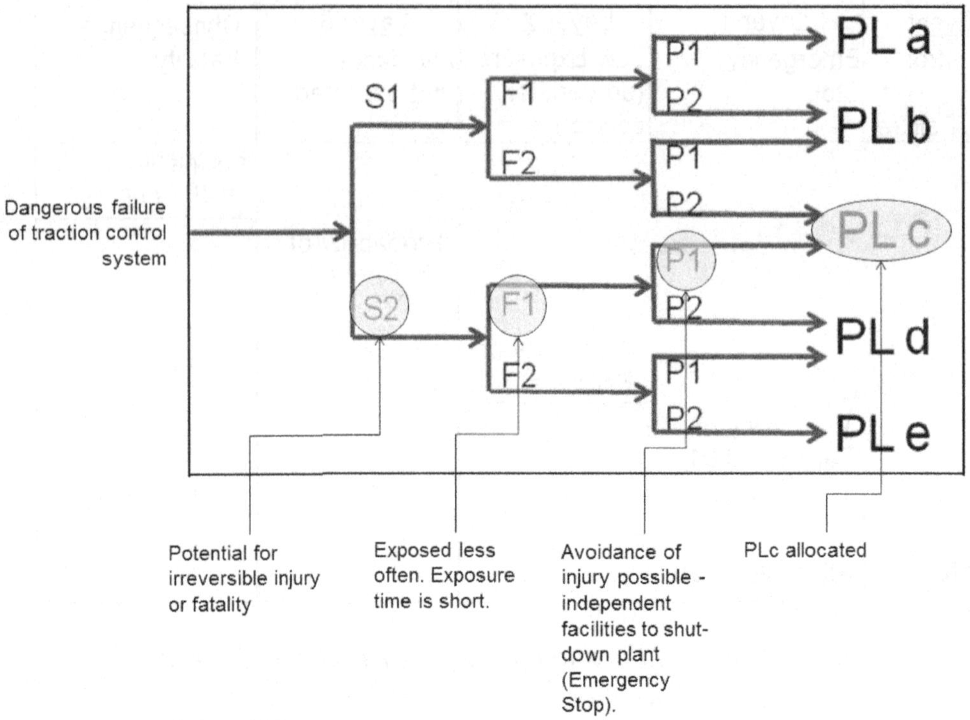

Given the complexity of mining and other plant, SIL and PL allocation should be performed via a quantitative or semi-quantitative methodology where practicable. Typically, project risk assessments yield a large number of hazards, many of which require further consideration and allocation of SIL's or PL's to safety-related systems. In such cases it may be impracticable to apply a semi-quantitative or quantitative approach due to the substantial amount of time and resources involved.

The alternative approach is to use a qualitative method as a "first pass" and then a semi-quantitative or quantitative method as a "second pass" for those hazards which:

- Pose a high EUC risk
- Have a safety-related system as the only risk control
- Involve safety-related systems which received a high SIL allocation (>SIL2) during the "first pass", or
- Did not achieve a satisfactory level of consensus within the multi-disciplinary team during the "first pass" which would justify a "second pass"

In the event that "tolerable risk" has not been specified by the end-user, has been stated to be 'zero', is otherwise unable to be determined, or where the responsibility for the SIL allocation activity has been delegated by the end-user to the designer, the use of the quantitative and semi-quantitative methods may prove fruitless. This is because quantitative SIL allocation methods require the definition of a numerical 'tolerable risk' frequency target in order to compute the risk reduction to be carried by each risk control or safeguard. Without this figure, the risk reduction cannot be computed and the SIL allocations cannot be determined. Section 7.6 provides a method for obtaining a concept of "tolerable risk" frequency if one has not been otherwise defined.

In the event that "tolerable risk" is not defined then the use of a qualitative SIL allocation may be justified by a designer, even for complex plant.

Product designers, manufacturers and suppliers are therefore left in a potentially challenging situation. The nature of the product cycle often requires that a designer needs to specify and implement safety functions in their products in order to get a mature product into the market-place. Often this may need to occur without reference to any end-user or customer specifications of what SIL's or PL's need to be achieved by those safety functions.

Typically, designers, manufacturers and suppliers may also intend to sell their products to numerous end-users, in numerous countries, all of which may have differing notions of "tolerable risk". How do they arrive at SIL or PL allocations that will meet the potential safety requirements of a broad range of future customers, and considering that the specific physical environment, procedures, and the full range of risk controls that will be used by each end-user are not known? A framework for solving this dilemma – by pre-emptive SIL / PL allocation - is provided in Section 8.2. In this case a successful outcome can be obtained by using a qualitative SIL or PL allocation method that is naturally conservative, such as one of the methodologies listed above.

7.8 AS(IEC)61508 Phase 9 - Safety Requirements Specification

Objective/s:

AS(IEC)61508 Phase 9 is primarily concerned with developing a specification for the overall safety-related systems in terms of the functional and safety integrity requirements of the safety functions that are to be implemented by it. Phase 9 should also include consideration of the "other technology" (ie. mechanical, pneumatic, hydraulic etc.) safety-related systems and "other risk reduction measures" (ie. guards, barriers, procedures etc.) in order to achieve the required functional safety.

Responsible Entity:

As with Phase 3, 4 and 5, the conduct of this phase should be the responsibility of the end-user of the EUC – eg. the mining company.

Activities to be Undertaken:

From AS(IEC)61508.1 Clause 7.10.2, AS(IEC)62061 Clause 5, AS(IEC)61511 Clause 10 and ISO13849-1 Clause 5 the following should occur:

1. The functional and safety integrity requirements of each safety function comprising the E/E/PE and "other technology" safety-related systems should be documented in a Safety Requirements Specification (SRS), which should form part of the overall specification of the EUC.
2. Any "other risk reduction measures" (ie. guards, barriers, procedures etc.) that are intended to work in co-ordination with the safety functions should be defined in the SRS.

Documentation to be Kept:

The information and results acquired should be maintained throughout the overall safety lifecycle. The Safety Lifecycle Dossier should be updated with the following information:

1. The Safety Requirements Specification (SRS) for the E/E/PE and "other technology" Safety-related Systems is completed during Phase 9, and should contain the information listed in AS(IEC)61508.1 Clause 7.10.2, AS(IEC)62061 Clause 5, AS(IEC)61511 Clause 10 or ISO13849-1 Clause 5 , as appropriate. For example, if AS(IEC62061 is to be used to develop the SRS, the following information should be provided:
a. A confirmation of which Safety Functions are to be implemented (in whole or in part) via E/E/PE technologies.
b. A confirmation of which safety functions are to be implemented (in whole or in part) via 'other technologies'.
c. A confirmation of the "other risk reduction measures" (ie. guards, barriers, procedures) which are required to be implemented in conjunction with the safety-related systems.
d. The expected environmental conditions and the general requirements on the safety-related systems in response to these conditions (eg. Dust, Moisture, Chemicals, Humidity, Temperature, Vibration, Shock, and Electro-magnetic interference).

e.	A definition of the "safe state" of the EUC.
f.	A statement of the required performance of each safety function, via the following:
i.	A name, identifier and description of each safety function
ii.	The required SIL or PL to be achieved by the safety function and the demand mode
iii.	A statement of the functional intent of each safety function
iv.	The frequency of operation / demands
v.	A description of the fault reaction required
vi.	The required response time of the safety function and its input / output devices
vii.	Interfaces to other systems or functions
viii.	The conditions or operating modes of the EUC under which the safety function will be active and disabled
ix.	The priority of those functions that can be simultaneously active, but can cause conflicting action in the EUC
x.	A description of the operating environment
xi.	Tests and associated facilities required
xii.	Rate of operating cycles, duty cycle, demand rate and utilisation

7.9 Low Demand or High Demand Mode? – A 'Common-Sense' Approach

The classification of safety functions as 'low demand', 'high demand' or 'continuous' mode of operation is a seemingly simple task according to AS(IEC)61508. Modes of operation are defined in AS(IEC)61508.4 as follows:

low demand mode *– where the safety function is only performed on demand, in order to transfer the EUC into a specified safe state, and where the frequency of demands is no greater than one per year.*

high demand mode *– where the safety function is only performed on demand, in order to transfer the EUC into a specified safe state, and where the frequency of demands is greater than one per year.*

continuous mode *- where the safety function retains the EUC in a safe state as part of normal operation.*

However, I believe this simplistic approach may be problematic when applied to mining plant and machinery and other applications in some circumstances[39]. Consider the following example.

An electrical sub-station is known to be susceptible to phase-to-phase faults on its output, with the potential for adverse safety consequences. A risk assessment has indicated that the fault may occur more than once per year, requiring a protective trip to isolate power, thereby preventing the adverse safety consequences.

According to AS(IEC)61508.4 this protective trip would be classified as a high demand safety function. However, specifying the protective trip as a high demand safety function may actually lead to an inadequate level of functional safety being achieved in the long term. Why? Well consider the following.

Suppose that the protective trip is specified and designed according to AS(IEC)61508 / AS(IEC)62061 as a high demand safety function. If a zero fault tolerant architecture (eg. '1-out-of-1', or no redundancy) is selected for the design, then the PFH calculation used to determine the SIL range achieved by the safety function is as follows[40]:

$$\boxed{PFH = \lambda_{du}}$$

where λ_{du} = dangerous undetected failure rate

The PFH would therefore be determined by the rate of dangerous undetected failure rates of the components used by the safety function, and no other factors. If the PFH of the components used in the protective trip is calculated to be $5x10^{-7}$ per hour, then the protective trip could be deemed to be capable of SIL2 performance for this 'high demand' application.

The problem with this approach is that the evaluation of the protective trip's probability of failure and safety integrity has been made without reference to a factor

[39] Similar views have been expressed by other authors. See pages 8-9 and Appendix 7 of Smith and Simpson, "*Functional Safety – A straightforward guide to applying IEC61508 and related standards*", 2nd Edition.

[40] See AS(IEC)62061 Clause 6.7.8.2 and AS(IEC)61508.6 Annex B.

which is known to significantly influence the likelihood that it will operate when phase-to-phase fault occurs – it's proof-testing / inspection interval.

Alternatively, a 'common-sense' approach would be to acknowledge that the protective trip is actually required to perform its function 'on demand', when the phase-to-phase fault occurs, rather than on a 'per hour' basis. In this case, the probability that the protective trip does not operate when the phase-to-phase fault occurs is given by its average Probability of Failure on Demand (PFD_{ave}). In this case the PFD_{ave} of the protective trip may be calculated according to the general equation for 1-out-of-1'low demand' safety functions[41].

$$PFD_{ave} = 1 - e^{-[(\lambda_{du}.(\frac{TI}{2}+MTTR)+\lambda_{dd}.MTTR]}$$

where λ_{du} = dangerous undetected failure rate

λ_{dd} = dangerous detected failure rate

MTTR = Mean Time To Repair

TI = Proof-test Interval

Importantly, if no proof-test interval is specified, the PFD_{ave} will eventually reach unity, meaning the protective trip should not be expected to operate when demanded. If the proof-testing interval for the protective trip is arbitrarily set at five (5) years the PFD_{ave} would be expected to be more than 0.01. This corresponds to SIL1 range for 'low demand' safety functions, not SIL2.

Clearly, proof-test intervals can significantly influence the reliability of safety functions operating 'on demand'. As such, the setting of proof-test intervals for these safety functions must be informed by the effect of the proof-test interval on the overall probability of failure of the safety function and whether the specified SIL can be achieved. This can only occur if an appropriate demand mode is selected for the safety function.

Therefore, to alleviate the potential for these issues, I prefer to use a 'common-sense' approach to the classification of safety functions, rather than rely upon the arbitrary demand rate boundary of one (1) year.

The 'common sense' approach is based on the actual definition of a safety function in AS(IEC)61508.4.

safety function *- function to be implemented by an E/E/PE safety-related system or other risk reduction measures, that is intended to* <u>*achieve*</u> *or* <u>*maintain*</u> *a safe state for the EUC, in respect of a specific hazardous event.*

The 'common sense' approach is based on the following steps:

1. Do an initial classification of a safety function on the basis of whether it <u>achieves</u> or <u>maintains</u> a safe state, then
2. Consider the anticipated demand rate.

[41] See AS(IEC)61508.6 Annex B.

A safety function which is used to achieve a safe state typically monitors for a certain condition (via sensors and logic) and then performs its allocated function (via logic and actuators) when that condition occurs. If such a safety function experiences a dangerous undetected failure, an adverse impact on safety would only occur in the event that a hazardous plant condition subsequently arises and the safety function is called upon. For example, if an Emergency Stop push-button fills up with rain-water it would not be expected to cause an accident. Instead, if left in this faulty state, it would limit an operator's ability to mitigate particular plant hazards that may arise in the future.

Such safety functions have been described in AS(IEC)61511.1 as 'demand mode safety instrumented functions', because the expectation on their failure probability is typically on a 'per demand' basis. Plant protective functions are typically used to achieve a safe state, and therefore usually operate in 'demand mode'.

Therefore, if a safety function is protective, or is used to achieve a safe state and its failure probability should logically be specified on a 'per demand' basis, then the 'low demand' PFD targets of AS(IEC)61508 should be applied at the appropriate SIL. However, if a 'demand mode' safety function is expected to be demanded more than once per year, I suggest that proof-tests are carried out at least twice as often as the anticipated demand rate on the safety function irrespective of whether the PFD calculations indicate a lower rate of proof-testing.

Conversely, when a safety function which is used to maintain a safe state fails, a hazard is expected to arise immediately or shortly thereafter. For example, if the traction control system or brakes fail on an item of mobile plant, the potential for an unplanned movement or vehicle collision arises immediately.

Such safety functions have been described in AS(IEC)61511.1 as 'continuous mode safety instrumented functions', because the expectation on their failure probability is typically on a 'per hour' basis. Plant control functions are typically used to maintain a safe state, and therefore usually operate in 'continuous' or 'high demand' mode.

Therefore, if the safety function is a control function, or is used to maintain a safe state and its failure probability should be specified on a 'per hour' basis, then the 'high demand' PFH targets of AS(IEC)61508 / AS(IEC)62061 should be applied at the appropriate SIL.

7.10 AS(IEC)61508 Phases 6 to 8 - Realisation Planning Phases

Objective/s:

AS(IEC)61508 Phases 6, 7 and 8 are, by definition, only concerned with developing plans for the implementation of E/E/PE safety-related systems. It is appropriate that this scope be extended to cover all elements involved in functional safety, ie:

- Installing and commissioning the E/E/PE and "other technology" safety-related systems and "other risk reduction measures" in a controlled manner
- Validating the E/E/PE and "other technology" safety-related systems and "other risk reduction measures"
- Operating and maintaining the E/E/PE and "other technology" safety-related systems and "other risk reduction measures"

Responsible Entity:

Whilst ultimate responsibility for ensuring completion of the requirements of these phases rests with the end-user (eg. the mining company), responsibility for conducting the activities may be delegated to the vendor or designer.

Activities to be Undertaken:

From AS(IEC)61508.1 Clauses 7.7.2, 7.8.2 and 7.9.2, the following plans should be developed. These plans should be broadened to include the "other technology" safety-related systems and "other risk reduction measures".

1. Installation and Commissioning Plan, including:
a. The installation schedule
b. Those responsible for the different parts of the installation
c. Procedures for the installation
d. The sequence in which the various elements are to be integrated
e. The criteria for declaring that all or parts of the E/E/PE safety-related systems ready for installation and for declaring installation activities complete
f. Procedures for resolution of failures and incompatibilities
g. The commissioning schedule
h. Those responsible for the different parts of the commissioning
i. Procedures for the commissioning
j. The relationships to the different steps in the installation
k. The relationships to the validation

2. Validation Plan, including:
a. Details of when the validation shall take place
b. Details of those who will perform the validation
c. Specification of the relevant modes of operation of the EUC and their relationships to the E/E/PE safety-related systems, "other technology" safety-related systems and "other risk reduction measures",
d. Specification of the E/E/PE safety-related systems, "other technology" safety-related systems and "other risk reduction measures", to be validated for each mode of EUC operation (before commissioning commences),
e. The technical strategy for the validation (eg. analytical methods, tests etc…),
f. The measures, techniques and procedures that shall be used for confirming that the allocation of safety functions has been carried out correctly (confirmation shall ensure that each safety function conforms to its functional and safety integrity requirements)
g. Specific reference to the achievement (or non-achievement) of each requirement specified in the Safety Requirements Specification (SRS)
h. The required environment in which the validation tests will take place
i. The pass and fail criteria
j. The policies and procedures for evaluating the results of the validation, particularly non-compliances
3. Operations and Maintenance Plan, including:
a. The routine actions which need to be carried out on the E/E/PE safety-related systems, "other technology" safety-related systems and "other risk reduction measures"
b. The actions and constraints that are necessary to prevent an unsafe state, to reduce demands on the E/E/PE safety-related systems, "other technology" safety-related systems and "other risk reduction measures" or to reduce the consequences of hazardous events
c. The documentation that needs to be maintained, showing the results of functional safety audits and tests
d. The documentation that needs to be maintained on hazardous incidents and all incidents with a potential to create a hazard event
e. The scope of the maintenance activities (as distinct from modification activities)
f. The actions to be taken in the event of hazards occurring
g. The contents to be included in the chronological documentation of the operations and maintenance activities

h.	The requirement that the routine maintenance activities which are carried out to detect unrevealed faults should be determined by a systematic analysis
i.	The requirement that the plan be agreed upon by those responsible for operating and maintaining the E/E/PE safety-related systems, "other technology" safety-related systems, "other risk reduction measures" and non-safety-related systems which have the potential to place demands upon the safety-related systems.

Documentation to be Kept:

The information and results acquired should be maintained throughout the overall safety lifecycle. The Safety Lifecycle Dossier should be updated with the following information:

1.	Installation and Commissioning Plan.
2.	Validation Plan.
3.	Operations and Maintenance Plan.

7.11 AS(IEC)61508 Phases 10 to 11 - Overall Concept for the Realisation Phases

Objective/s:

The design and development of the E/E/PE and "other technology" safety-related systems and "other risk reduction measures" must follow clearly-defined processes, be subject to verification and be thoroughly planned and documented.

Responsible Entity:

Whilst ultimate responsibility for ensuring completion of the requirements of these phases rests with the end-user (eg. the mining company), responsibility for conducting the activities may be delegated to the vendor or designer.

Activities to be Undertaken:

It is suggested that the following should be used as an overall framework for the conduct of the realisation phases.

1. Design and development of the E/E/PE safety-related systems should follow the processes provided by AS(IEC)61508.2, AS(IEC)62061 Figure 2, Clause 6 and Annex B, AS(IEC)61511.2 or ISO13849-1 Figure 3 and Clause 4.4, as appropriate. Note: There is no standard terminology for describing safety functions and safety-related systems in the functional safety and machinery safety standards[42]. For example, in AS(IEC)62061 "safety functions" are referred to as "Safety-related Control Functions (SRCF's) and "E/E/PE safety-related systems" are referred to as "Safety-related Electrical Control Systems" (SRECS).
2. Design and development of "other technology" (ie. mechanical, pneumatic, hydraulic etc…) safety-related systems should follow the process provided by ISO-13849-1 Clause 4.4 and Figure 3. These systems are not within the scope of AS(IEC)61508, AS(IEC)62061 or AS(IEC)61511, however in AS(IEC)62061 reference is made to ISO13849-1 and ISO13849-2 for design and development of these systems.
3. Where a regulation, code of practice or standard exists for certain parts of a safety-related system (eg. fixed gas detection systems, remote control systems, conveyor remote isolation systems, Emergency Stops for plant etc…), then this should take precedence for the design and development of those parts.
4. Design and development of "other risk reduction measures" should follow the requirements of the regulations, codes of practice, standards, or guidelines applicable to the application (eg. AS4024.1 Part 1601 for fixed or movable guards).

Documentation to be Kept:

See Sections 7.12 to 7.19 and Appendix B.

[42] See Appendix A for details of the terminology used by each standard.

7.12 AS(IEC)61508 Phase 10 - E/E/PE Safety-related Systems Realisation (using AS(IEC)62061)

Objective/s:

The design and development of the E/E/PE safety-related systems should follow a ten (10) step process, as provided by AS(IEC)62061 Figure 2 (shown below in Figure 7.6). Guidance on the implementation of this process is provided in AS(IEC)62061 Clause 6.6.2 and AS(IEC)62061 Annex B.

Figure 7.6: AS(IEC)62061 Design and Development Process

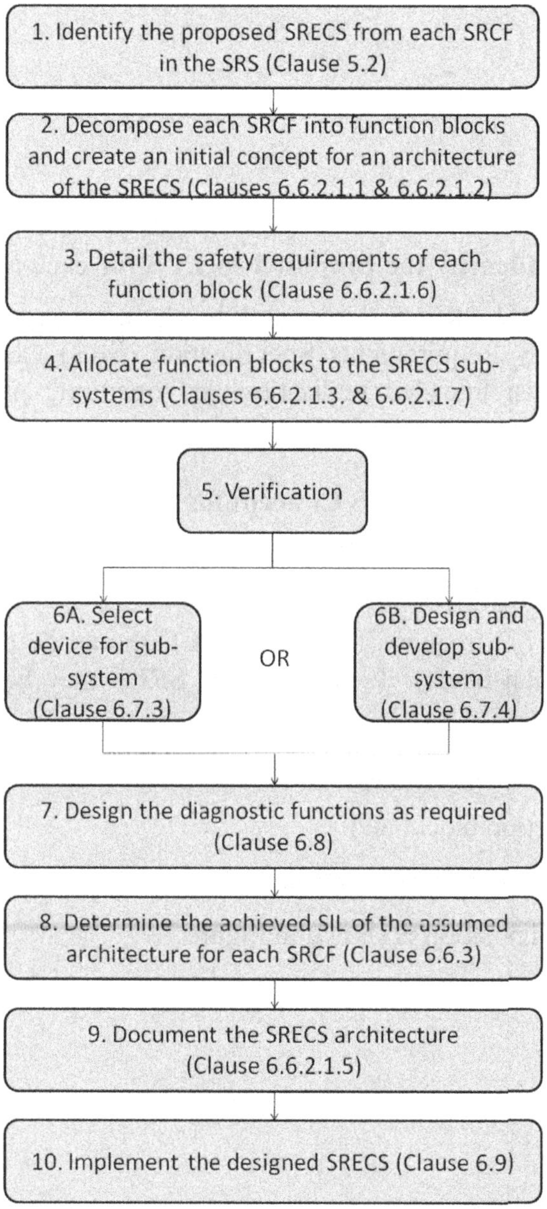

Responsible Entity:

Whilst ultimate responsibility for ensuring completion of the requirements of these phases rests with the end-user (eg. the mining company), responsibility for conducting the activities may be delegated to the vendor or designer.

Activities to be Undertaken / Documentation to be Kept:

This section provides a worked example of how the design and development steps are applied to a set of SRCF's that have been identified for an item of plant. An item of mobile mining plant is the subject (eg. shuttle cars, continuous miners etc…), and the SRCF's that have been defined for the purposes of the worked example (among others that would normally be defined) are as follows:

1. Traction Control, including:
    a. Start / Stop
    b. Forward / Reverse
    c. Fast / Slow
2. Service Brakes
3. Steering
4. Emergency Stop
5. Pre-start Alarm
6. Tramming Alarm

AS(IEC)62061 Step 1 – Identify the Proposed SRECS for each SRCF in the SRS.

Step 1 of the AS(IEC)62061 SRECS design and development process is achieved by the drafting of the Safety Requirements Specification (SRS) where the functional intent, inputs and outputs, intended realisation and operating parameters of each SRCF are documented along with the required SIL.

AS(IEC)62061 Step 2 – Initial SRECS Concept for the SRCF's

The objective is to decompose each SRCF into function blocks to create an initial concept for how they will be implemented in the SRECS.

AS(IEC)62061 Clause 6.6.2, requires that each SRCF should be decomposed into a structure of SRCF function blocks. Typically, each SRCF may be decomposed into three (3) function blocks:

- a sensing / initiation function block
- a logic / resolver function block, and
- an actuation / final element function block

An example of this decomposition is shown in Figure 7.7.

Figure 7.7: AS(IEC)62061 Step 2

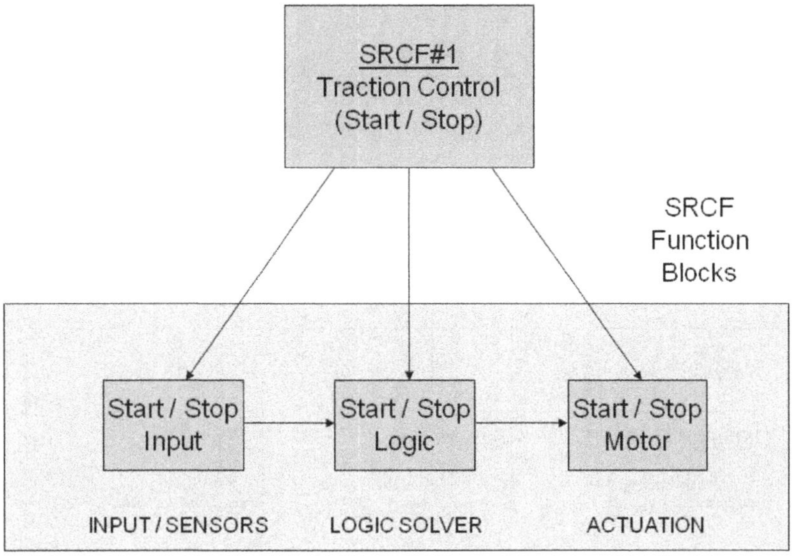

A 'SRECS Concept Design' document should be initiated with the following information, and a copy should be provided by the designer for inclusion in the Safety Lifecycle Dossier.

- A graphical representation of each SRCF and its SRCF functional blocks.
- A definition of the inputs and outputs of each SRCF function block.

AS(IEC)62061 Step 3 – Detail the Safety Requirements of the SRCF Functional Blocks

The objective is to specify the safety requirements of each of the SRCF function blocks used by the SRCF's.

From AS(IEC)62061 Clause 6.6.2.1.6 the following should occur:

1. The safety requirements of each SRCF function block should be as specified for the parent SRCF in the SRS.
2. These safety requirements include the functional requirements (eg. input information, internal logic operation and output of the function block), and the safety integrity requirements (ie. SIL).
3. If any SRCF functional block is to be utilised by more than one SRCF, then the safety requirements of that SRCF functional block should be the highest of all those SRCF's it has been allocated to.

An example of SRCF function block allocation is shown in Figure 7.8.

Figure 7.8: AS(IEC)62061 Step 3

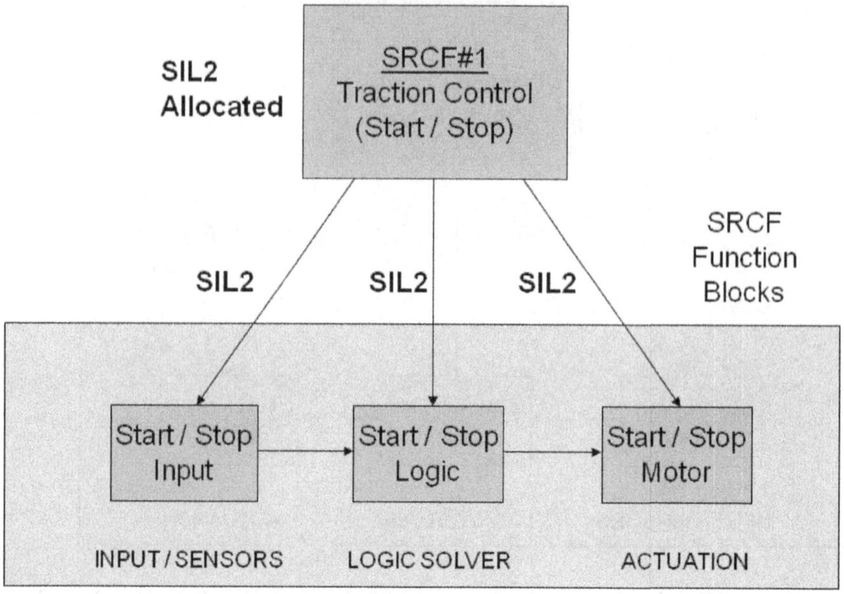

The SRECS Concept Design document should be updated with the following information, and a copy should be provided by the designer for inclusion in the Safety Lifecycle Dossier.

1. A graphical representation of the SRCF's into functional blocks.
2. A definition of the inputs and outputs of each SRCF function block.
3. The safety requirements (ie. functional and safety integrity) for each SRCF function block.

AS(IEC)62061 Step 4 –Allocate SRCF Function Blocks to SRECS Sub-systems

The objective is to create an initial concept for the physical implementation of the SRCF's by the SRECS sub-systems.

From AS(IEC)62061 Clause 6.6.2 the following should occur:

1. Each SRCF function block should be allocated to a SRECS sub-system within the architecture of the SRECS.
2. More than one SRCF function block may be allocated to each SRECS sub-system.
3. The safety requirements for the SRCF sub-systems should be those of the SRCF function blocks allocated to them.
4. If more than one SRCF functional block is allocated to a sub-system, then the highest integrity requirement applies.
5. Where software is to be utilised by SRECS sub-systems which implement the SRCF function blocks, the requirement for a Software Safety Requirements Specification should be identified.

The SRCF function block to SRECS sub-system allocation process is shown in AS(IEC)62061, as shown in Figure 7.9.

Figure 7.9: AS(IEC)62061 Step 4

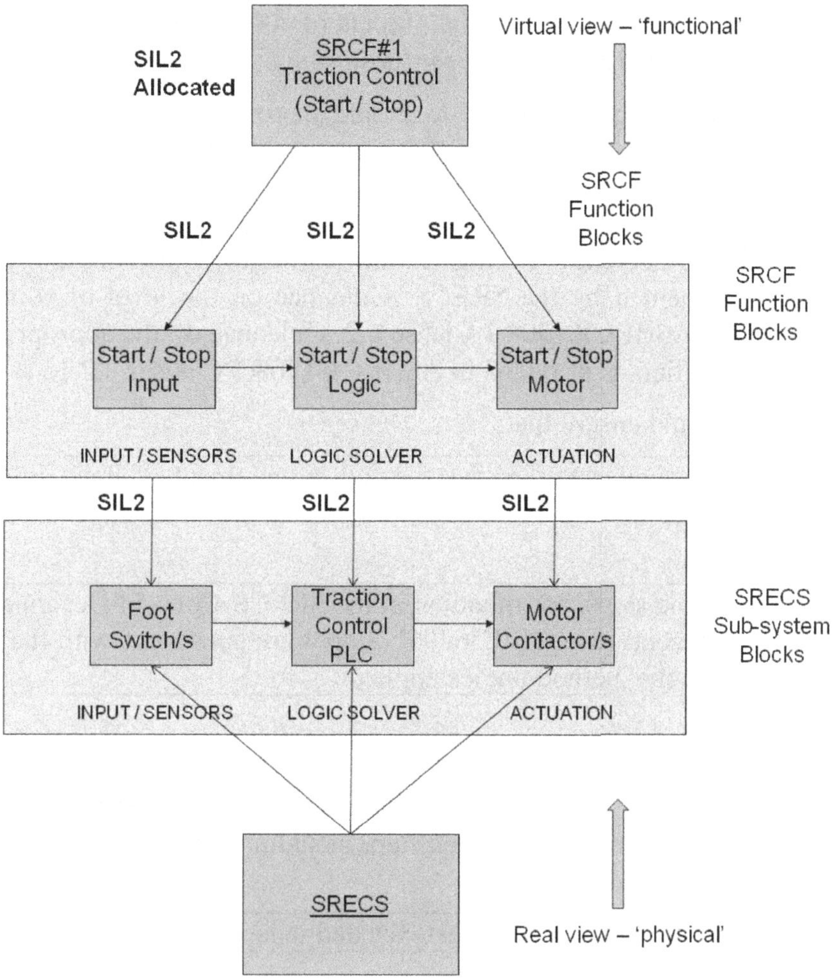

The SRECS Concept Design document should be updated with the following information, and a copy should be provided by the designer for inclusion in the Safety Lifecycle Dossier.

1. A graphical representation of the SRCF, SRCF functional block and SRECS sub-system architecture.
2. A definition of the inputs and outputs of each SRCF function block and SRECS sub-system.
3. The safety requirements (ie. functional and safety integrity) allocations for each SRCF function block and SRECS sub-system.
4. Identification of those SRCF function blocks which have been allocated to SRECS sub-systems that will involve a complement of software.
5. A Sub-system Safety Requirements Specification. If the principal SRECS designer intends to delegate or sub-contract the design of any SRECS sub-system to another organisation, they should formally communicate the SRECS sub-system safety requirements to the sub-contractor / delegate in a "Sub-system Safety Requirements Specification".

At the conclusion of Step 4, the SRECS Concept Design document should be complete enough for use by the designers during the detailed design.

AS(IEC)62061 Step 5 –Verification

The objective is to demonstrate that the outputs of AS(IEC)62061 Steps 1 to 4 fully satisfy the objectives and requirements for those steps.

This verification should be conducted to ensure that the concept architecture of the SRECS, and the decomposition into SRCF function blocks and SRECS sub-systems and the safety requirements allocations are correct, and that the requirements of the SRS can be met. The verification should be performed by a person with a level of independence and competence, commensurate with that required by the highest SIL SRCF being implemented by the SRECS. Guidance on the level of competence of verifiers is given in AS(IEC)61508.1 Clause 6.2. Guidance on the appropriate level of independence of verifiers is provided in AS(IEC)61508.1 Clauses 8.2.15 to 8.2.18.

This verification should ensure that:

1. The Safety Lifecycle Dossier has the full complement of input information, analysis and design documentation and results for this stage of the safety lifecycle.
2. The results of the steps, as embodied in the SRECS Concept Design document produced at the end of AS(IEC)62061 Step 4 are consistent with the inputs to the process and the methodologies applied.

The Safety Lifecycle Dossier should be updated with the following information:

1. Verification Report. This should comprise an audit checklist based on the objectives and requirements of each step, including observations made and corrective actions required.
2. A statement of the verifier's competence and independence.

AS(IEC)62061 Step 6 –Design and Develop SRECS Sub-systems

The objective is to realise SRECS sub-systems that fulfill all of the safety requirements of the allocated SRCF function blocks for each SRCF.

From AS(IEC)62061 Clause 6.7.3 and 6.7.4 the following should occur:

1. The Hazard and Risk Analysis, Safety Requirements Allocation, Safety Requirements Specification and SRECS Concept Design document should be provided to the SRECS designers,
2. The design of each SRECS sub-system should comply with the SRS,
3. Each SRECS sub-system should be realised in order to fulfill the functional and safety integrity requirements of all SRCF function blocks allocated to it.
4. If software is to be incorporated into any SRECS sub-system, then the requirements of AS(IEC)62061 Clauses 10, 11 and 12 and Annex C and AS(IEC)61508.3 should be followed, as applicable to the SIL allocated to those SRECS sub-systems.
5. A 'Software Safety Requirements Specification' should be produced for SRECS software according to AS(IEC)62061 Clause 10.

An iterative design process is suggested for the design and development of each SRECS sub-system. The requirements for the design and development of hardware and software used in SRECS sub-systems are covered in AS(IEC)62061 Clauses 6.7, 6.9 and 6.11.

The key focus of the design must be on the following aspects:

1. Satisfaction of the architectural constraints for the applicable SIL for each SRECS sub-system involved in performing the SRCF's. This will usually involve the calculation or sourcing of Safe Failure Fraction (SFF) and dangerous failure rate values for the components / modules to be used,
2. Achieving either a total Probability of Failure on Demand (PFD) or Probability of Failure per Hour (PFH) value for each SRCF, as applicable, that complies with the target failure values for the SIL that was allocated to the SCRF. This may involve an iterative reliability analysis to ensure that the detailed design and component selections are sufficient to achieve the target SIL.
3. Achieving the requirements for avoidance and control of systematic failures for both hardware and software used in the SRECS. These requirements are detailed in AS(IEC)62061 Clauses 6.4, 6.7.9, 6.10, 6.11, 6.12 and Annex C, AS(IEC)61508.2 and AS(IEC)61508.3 Annex A. This will involve the gathering of evidence of activities undertaken during the specification, design and implementation stages in order to make a deterministic argument that the systematic failure requirements of the target SIL have been met.

AS(IEC)62061 Figure 4 provides a description of the SRECS sub-system design and development process, as shown in Figure 7.10.

Figure 7.10: AS(IEC)62061 Step 6

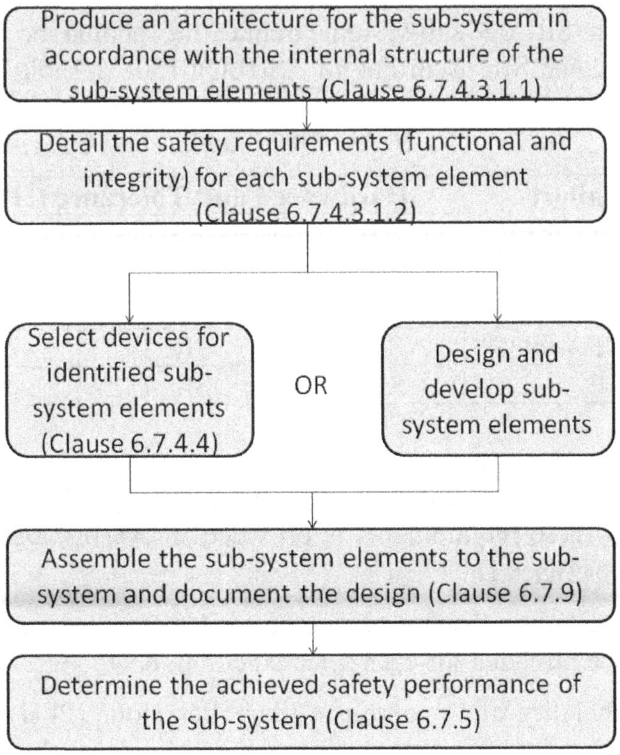

Two approaches to SRECS sub-system design may be considered:

1. Selection of devices that have already been proven to fulfill the allocated safety requirements for the SRECS sub-system.
2. Design and development of SRECS sub-systems by combining devices which have not previously been proven to fulfill the allocated safety requirements, but specifying how they are to be arranged and interact, in order that the allocated safety requirements may be achieved.

If the former approach (selection of proven devices) is undertaken by the SRECS designer, then the following requirements should apply:

1. Documentation proving compliance to AS(IEC)61508 for the allocated SIL (eg. functional safety certificates) should be provided for each selected component.
2. Documentation detailing the Safe Failure Fraction (SFF) and Hardware Fault Tolerance (HWFT) should be provided for each selected component.
3. If the selected devices are Type A, as defined in AS(IEC)61508.2 Clause 7.4.3.1.2, then a SRECS sub-system architecture should be selected that will achieve the allocated SIL according to AS(IEC)61508.2 Table 2 (shown in Table 7.3a).

Table 7.3a: Architectural Requirements (Type A)

Safe Failure Fraction (SFF)	**Hardware Fault Tolerance (HWFT)**		
	0	**1**	**2**
SFF < 60%	SIL 1	SIL 2	SIL 3
60% ≤ SFF < 90%	SIL 2	SIL 3	SIL 4
90% ≤ SFF < 99%	SIL 3	SIL 4	SIL 4
99% ≤ SFF	SIL 3	SIL 4	SIL 4

4. If the selected devices are Type B, as defined in AS(IEC)61508.2 Clause 7.4.3.1.3, then a SRECS sub-system architecture should be selected that will achieve the allocated SIL according to AS(IEC)61508.2 Table 3 (shown in Table 7.3b).

Table 7.3b: Architectural Requirements (Type B)

Safe Failure Fraction (SFF)	**Hardware Fault Tolerance (HWFT)**		
	0	**1**	**2**
SFF < 60%	Not allowed	SIL 1	SIL 2
60% ≤ SFF < 90%	SIL 1	SIL 2	SIL 3
90% ≤ SFF < 99%	SIL 2	SIL 3	SIL 4
99% ≤ SFF	SIL 3	SIL 4	SIL 4

5. If the selected devices contain embedded software then the design of the software should comply with the requirements of AS(IEC)61508.3. A guideline for complying with these requirements is provided in AS(IEC)62061 Annex C and AS(IEC)61508.6 Annex D.
6. Documentation detailing the rate of dangerous detected and dangerous undetected failures should be provided for each selected component.
7. An overall Probability of Dangerous Failure Per Hour (PFH) or Probability of Failure on Demand (PFD), as appropriate, should be determined for each SRECS

sub-system and SRCF, which must take account of the selected architecture of the SRECS sub-systems, the potential for Common Cause Failures (CCF) affecting redundant components, testing intervals, diagnostic coverage of the dangerous failures, repair times, etc. This calculation should be performed according to AS(IEC)62061 Clause 6.7.8.2 or AS(IEC)61508.6 Annex B, as appropriate.

8. Appropriate proof-testing intervals must be selected for each SRECS sub-system such that an appropriate PFH or PFD can be achieved for the target SIL. It is crucial that these proof-testing intervals are implemented during the operation and maintenance phase, otherwise the intended level of safety integrity may not be achieved.

9. If any of the selected devices have previously been proven to be designed to a Performance Level (PL) according to ISO13849-1 then the equivalent SIL of those devices should be determined and documented according to ISO13849-1 Table 4, (shown in Table 7.4).

Table 7.4: AS(IEC)61508 SIL and ISO13849-1 Equivalences

Performance Level (PL)	Equivalent Safety Integrity Level (SIL)
a	No SIL (or <SIL1)
b	SIL 1
c	SIL 1
d	SIL 2
e	SIL 3

If the latter approach (design and development of SRECS sub-systems) is undertaken, then the following requirements should apply:

1. The SFF and HWFT of the proposed SRECS sub-system should be determined via a reliability analysis. The analysis should determine all relevant failure modes of the proposed SRECS sub-system components and interfaces, whether each failure is a single point of failure, is potentially safe or dangerous, or will be detectible or undetectable. The SFF calculation should be conducted according to AS(IEC)62061 Clause 6.7.7, and with reference to AS(IEC)61508.2 Annex C and AS(IEC)61508.6 Annex C. The calculation may take advantage of dependable failure rate data collected from field experience by manufacturers or other users which is relevant to the intended use, component failure data from a recognised industry data source which is relevant to the intended use, failure mode data given in AS(IEC)62061 Annex D or failure rate data derived from the results of testing and analysis.

2. If the proposed SRECS sub-system uses Type A devices as defined in AS(IEC)61508.2 Clause 7.4.3.1.2, then a SRECS sub-system architecture should be selected that will achieve the allocated SIL according to AS(IEC)61508.2 Table 2 (shown in Table 7.3a).

3. If the proposed SRECS sub-system uses Type B devices as defined in AS(IEC)615080.2 Clause 7.4.3.1.3, then a SRECS sub-system architecture should be selected that will achieve the allocated SIL according to AS(IEC)61508.2 Table 3 (shown in Table 7.3b).

4. If the selected devices contain embedded software then the design of the software should comply with the requirements of AS(IEC)61508.3. A guideline for complying with these requirements is provided in AS(IEC)62061 Annex C and AS(IEC)61508.6 Annex D.
5. An overall Probability of Dangerous Failure Per Hour (PFH) or Probability of Failure on Demand (PFD), as appropriate, should be determined for each SRECS sub-system and SRCF, which must take account of the selected architecture of the SRECS sub-systems, the potential for Common Cause Failures (CCF) affecting redundant components, testing intervals, diagnostic coverage of the dangerous failures, repair times, etc. This calculation should be performed according to AS(IEC)62061 Clause 6.7.8.2 or AS(IEC)61508-6 Annex B, as appropriate.
6. Appropriate proof-testing intervals must be selected for each SRECS sub-system such that an appropriate PFH or PFD can be achieved for the target SIL. It is crucial that these proof-testing intervals are implemented during the operation and maintenance phase, otherwise the intended level of safety integrity may not be achieved.

A dossier of documentation associated with the design process of each of the SRECS sub-systems should be provided by the designer for inclusion in the Safety Lifecycle Dossier. The required information is specified in AS(IEC)62061 Clause 6.7.

1. SRECS sub-system functional specifications for the functions and interfaces which will be used by the SRCF's,
2. SRECS sub-system realisation (actual design drawings, assemblies and component selections),
3. Component data sheets,
4. The results of Factory Acceptance Testing,
5. Component functional safety certifications and other certifications, such as EMI compliance, Ex-ratings etc..,
6. Calculations of Hardware Fault Tolerance (HWFT), Safe Failure Fraction (SFF) and either PFD or PFH values for each SRECS sub-system and SRCF, as appropriate to the demand mode.
7. Component / module reliability data to support the above calculations, including estimated rates of failure of any failure modes that would cause a dangerous failure of the SRECS sub-systems,
8. A statement of the fault exclusions claimed when estimating the HWFT and SFF figures,
9. Probability of dangerous transmission errors for digital data communication processes, where applicable.
10. Constraints due to environmental or operating conditions and the lifetime of the sub-system which should not be exceeded,
11. Test and maintenance requirements,
12. Diagnostic coverage and diagnostic testing interval,
13. Information to allow the calculation of a Mean Time to Repair (MTTR),

14. Sub-system Type (A or B), SFF, HWFT and SIL Claim Limit of each SRECS sub-system,
15. Limits on the application of the SRECS sub-system,
16. The highest SIL that can be claimed by a SRCF which uses the SRECS sub-system, on the basis of:
a. Measures taken against the introduction of systematic faults during design and implementation of hardware and software,
b. Design features which make the sub-system tolerant of systematic faults,
17. Information to enable hardware and software configuration management.

AS(IEC)62061 Step 7 –Design and Develop Diagnostics

The objective is to realise SRECS diagnostic functions that are necessary to fulfill the requirements of the architectural constraints and safety integrity requirements of the allocated function blocks of the SRCF's.

From AS(IEC)62061 Clause 6.8 the following should occur:

1. The diagnostic functions should be considered as separate functions that may have a different structure to the SRCF's and may be performed by:
a. The same sub-system which requires the diagnostics (ie. embedded diagnostics in the sensor devices),
b. Other sub-systems of the SRECS (eg. PLC-based diagnostics of the sensors),
c. Non-SREC sub-systems.
2. Diagnostics should be designed to satisfy certain requirements that are also applicable to the SRCF's, namely:
a. Requirements for avoidance of systematic failure,
b. Requirements for control of systematic failures.
3. The probability of dangerous failure of the SRECS diagnostic functions should be taken into account when estimating the probability of dangerous failure of the SRCF's.
4. A clear description of the SRECS diagnostic functions, their failure detection and an analysis of their contribution to the safety integrity of the SRCF's (ie. probability of dangerous failure) should be provided.

An iterative design process is also suggested for the design of SRECS diagnostics. The requirements for the design and development of diagnostics utilised in SRECS sub-systems are covered in AS(IEC)62061 Clause 6.8. The selection of diagnostic capabilities may be steered by the reliability analysis conducted during AS(IEC) 62061 Step 6, which identifies those dangerous failures which should be detectable.

The key focus of the design of diagnostics must be to maximise the proportion of safe and dangerous failure modes that are detectible, as this will have the following effect upon the safety integrity of the SRECS sub-systems:

- It will increase the Diagnostic Coverage (DC) of the devices / modules
- It will increase the SFF of the devices / modules, allowing a higher architectural SIL claim limit for a given Sub-system Type and HWFT
- It will also decrease the proportion of SRECS sub-system failure modes that are dangerous and undetected, which will reduce the predicted PFH and PFD values of the SRECS sub-systems

Suppliers of proven devices may be able to provide information on diagnostic coverage from the independent functional safety certifications that have been obtained for their components. Such information should be obtained where possible as it may allow the selection of simpler SRECS sub-system architectures.

An example of SRECS-subsystem diagnostics is shown in Figure 7.11 below.

Figure 7.11: Options for SRECS Sub-systems Diagnostics

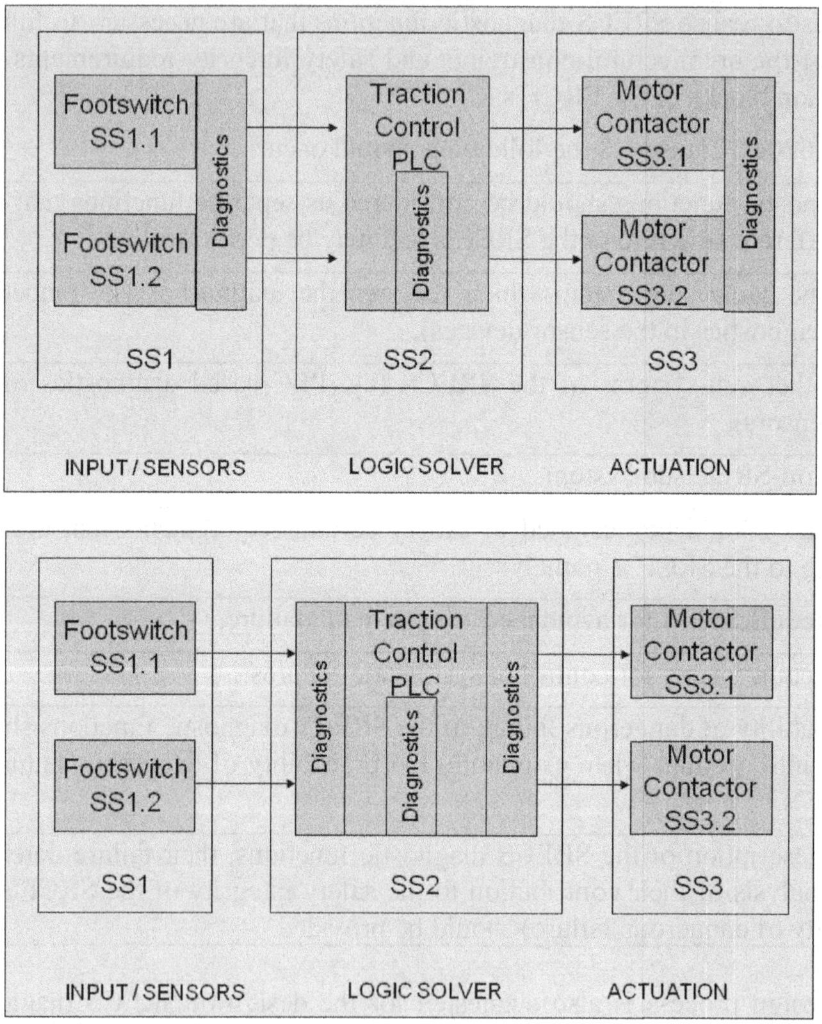

A dossier of documentation associated with the design process of each of the SRECS sub-systems should be provided by the designer for inclusion in the Safety Lifecycle Dossier. This should include:

1. SRECS Diagnostics Concept document which apportions diagnostic functions to each of the SRECS sub-systems.	
2. SRECS Diagnostics Specification for each SRECS sub-system (this may be included with the SRECS sub-system design documentation provided in Step 6), including:	
	a. Clear description of the diagnostic functions
	b. Failure modes covered
	c. Method of detection of failures
	d. Diagnostic coverage expected to be achieved
	e. Reaction of the diagnostics upon detection of a failure
	f. Probability of failure of the diagnostic function (so that its contribution to the overall safety integrity of the associated SRCF's can be ascertained)

AS(IEC)62061 Step 8 –Determine the Achieved SIL for each SRCF

The objectives are to:

- Determine, via a reliability analysis, that the target failure measures for the allocated SIL for each SRCF, have been met by the SRECS
- Confirm that the architectural constraints applicable to the allocated SIL for each SRCF, have been met by the SRECS.

This step itself comprises a verification of AS(IEC)62061 Steps 6 and 7.

From AS(IEC)62061 Clause 6.6.3 (using the information sourced in Steps 4, 6 and 7), the following should occur:

1. The SIL that can be achieved by the SRECS should be considered separately for each SRCF to be performed by the SRECS,
2. The PFD or PFH, as appropriate, for each SRCF should be calculated and shown to be equal to or less that the target PFD or PFH for the SIL specified for the SRCF in the SRS. This calculation should take into account the architecture of the SRECS as it relates to the SRCF under consideration and the estimated rate of failure of each SRECS sub-system used to perform the SRCF.
3. The SIL Claim Limit (SILCL) achieved by a SRCF according to the architectural constraints should be less than or equal to the lowest SILCL of any SRECS sub-system involved in the performance of the SRCF.

Verification of the SIL of each SRCF should be carried out by persons will an appropriate degree of independence and competence. Guidance on the level of competence of verifiers is given in AS(IEC)61508.1 Annex B. Guidance on the appropriate level of independence of verifiers is provided in AS(IEC)61508.1 Clause 8.2.14.

The verification process, regardless of the SIL allocation, must include a confirmation of the following features of each SRCF:

1. Satisfaction of the architectural constraints on SFF, HWFT and Sub-system Type for the applicable SIL for each SRECS sub-system involved in performing the SRCF's.
2. Achieving either a total PFD or PFH value for each SRCF, as applicable, that complies with the target failure values for the SIL that was allocated to the SCRF in the SRS. These values are provided AS(IEC)61508.1 Tables 2 and 3 (shown in Table 7.5).

Table 7.5: PFH / PFD Requirements

Safety Integrity Level (SIL)	Low Demand Operation Probability of Failure on Demand (PFD) (average probability of failure to perform its design function on demand)
1	$10^{-2} \leq PFD < 10^{-1}$
2	$10^{-3} \leq PFD < 10^{-2}$
3	$10^{-4} \leq PFD < 10^{-3}$
4	$10^{-5} \leq PFD < 10^{-4}$

Safety Integrity Level (SIL)	High Demand or Continuous Operation Probability of Dangerous Failure per Hour (PFH) (probability of a dangerous failure per hour)
1	$10^{-6} \leq PFH < 10^{-5}$
2	$10^{-7} \leq PFH < 10^{-6}$
3	$10^{-8} \leq PFH < 10^{-7}$
4	$10^{-9} \leq PFH < 10^{-8}$

3. Achieving the requirements for avoidance and control of systematic failures for both hardware and software used in the SRECS, according to the requirements of AS(IEC)62061 Clauses 6.7.9, 6.11 and Annex C and AS(IEC)61508.2 and AS(IEC)61508.3.

An example of a SRCF SIL calculation using the Fault Tree Analysis (FTA) techniques is shown in Figure 7.12 below. This diagram shows the SRECS sub-systems involved in the SRCF, the overall PFH result and the contribution of each SRECS sub-system to the overall PFH.

Figure 7.12: SIL Calculation Using Fault Tree Analysis (FTA)

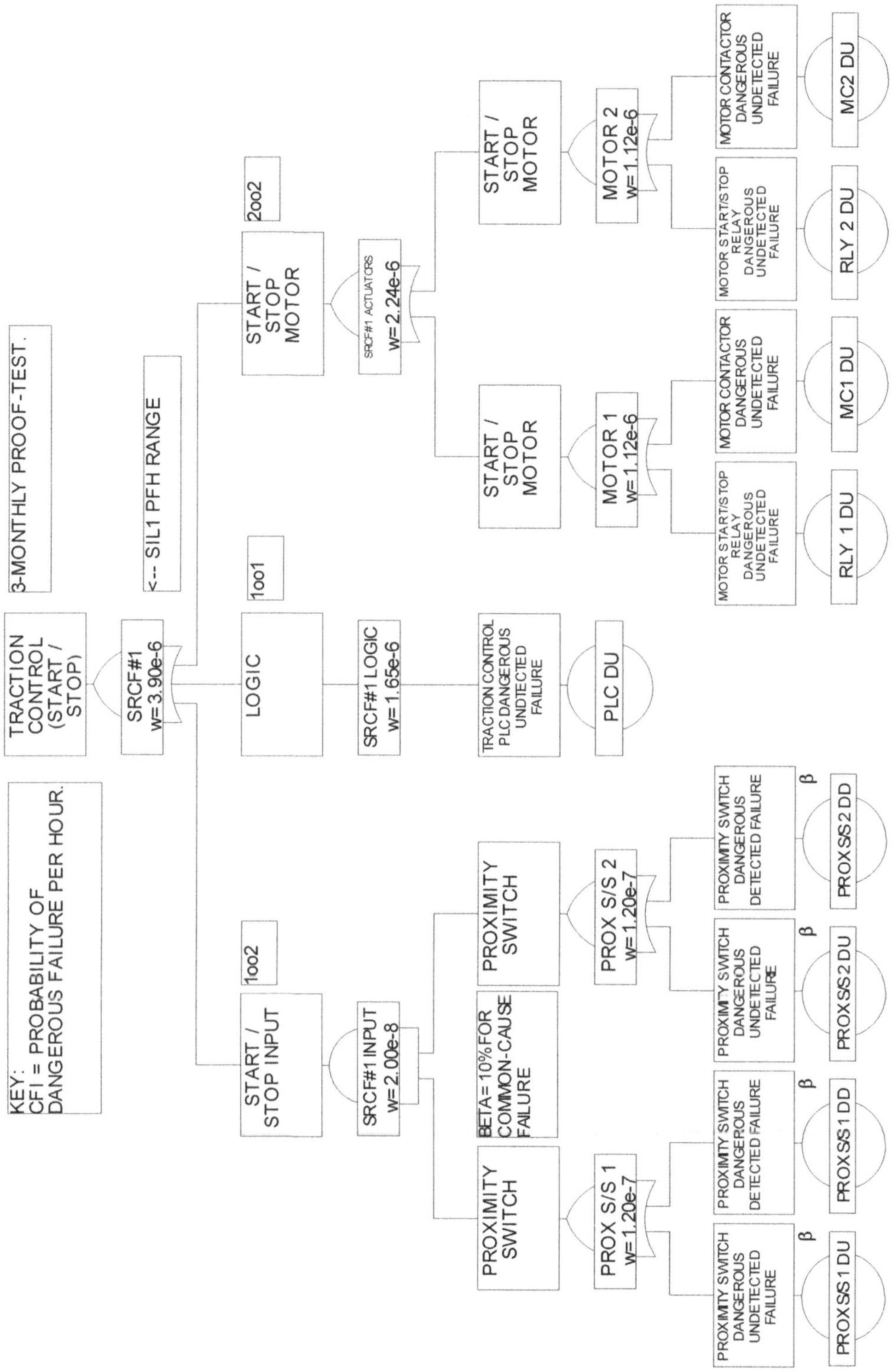

The Safety Lifecycle Dossier should be updated with the following information:

1. A SIL Verification Report addressing the aforementioned features of each SRCF.
2. A statement of the verifier's competence and independence.

AS(IEC)62061 Step 9 –Document the SRECS Architecture

The objective is to provide a comprehensive overview of all SRCF's, their contributing SRCF function blocks, and the corresponding SRECS sub-systems implementing them within the SRECS.

From AS(IEC)62061 Clause 6.6.2.1.5 the architecture of the SRCF's, function blocks and SRECS sub-systems should be documented describing the overall allocations and the functional and physical inter-relationships.

A SRECS Architecture Diagram should be provided by the designer for inclusion in the Safety Lifecycle Dossier, including.

1. A graphical representation of the SRCF, SRCF functional block and SRECS sub-system architecture.
2. A definition of the inputs and outputs of each SRCF function block and SRECS sub-system.
3. The safety requirements (ie. functional and safety integrity) allocations for each SRCF function block and SRECS sub-system.
4. Identification of those SRCF function blocks which have been allocated to SRECS sub-systems that will involve a complement of software.
5. Identification of the SRECS sub-system internal architecture selected (ie., SFF, HWFT and Sub-system Type) and the PFD / PFH and SILCL achieved by each.
6. Identification of the allocation of diagnostic functions to be performed within the SRECS sub-systems, or by other sub-systems.

An example of SRECS Architecture diagram is shown in Figure 7.13.

Figure 7.13: SRECS Architecture Diagram

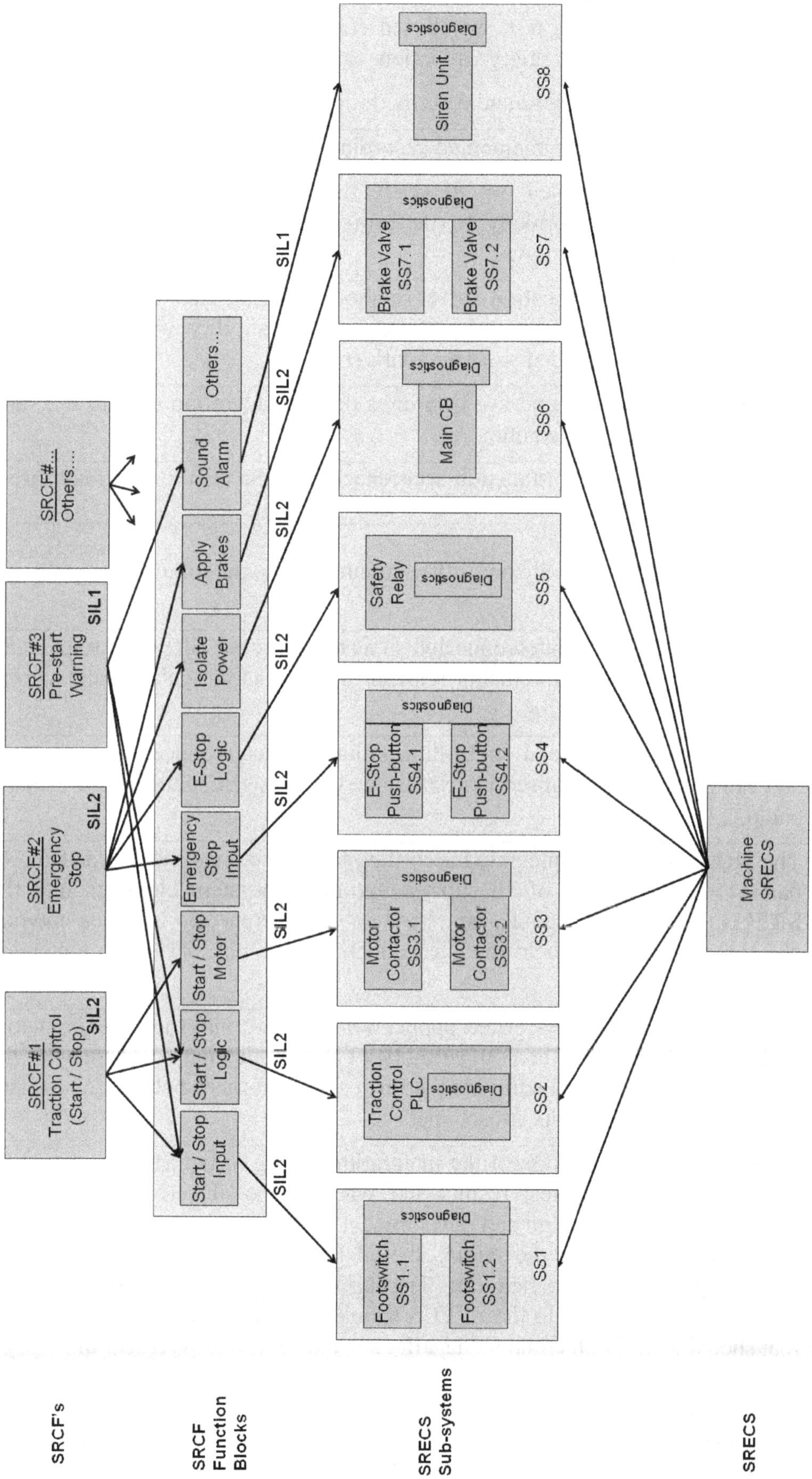

AS(IEC)62061 Step 10 –Implement the Designed SRECS

The objective is to integrate, test, install and commission the SRECS so that it is ready for intended use and for safety validation.

From AS(IEC)62061 Clause 6.12 and 6.13 the following should occur:

1. The SRECS should be implemented according to the specified design,
2. Testing should be carried out to verify that SRECS sub-system elements behave and interact correctly to perform their intended function and not perform unintended functions,
3. The integration of application software should include tests specified during the design and development to ensure its compatibility with the SRECS hardware and its embedded software platform,
4. Testing to reveal faults and avoid failures due to integration of hardware and software should be carried out,
5. The SRECS should be installed in accordance with the plan for the final safety validation.

For SRECS sub-system hardware and software integration, the following should apply:

1. The SRECS should be interconnected so as to satisfy the SRECS Architecture Diagram and those requirements relevant to conductors, cabling and wiring practices given in AS60204.
2. Measures for avoiding and controlling failures of interconnecting conductors and cables should be realised in accordance with AS(IEC)62061 Clauses 6.4.1 and 6.4.2.
3. The SRECS should be integrated according to the specified SRECS design. As part of the integration of all subsystems and subsystem elements into the SRECS, the SRECS should be tested to verify that all modules interact correctly to perform their intended function and not perform unintended functions.
4. The integration of safety-related application software into the SRECS should include tests to ensure the compatibility of the application software with the hardware and embedded software platform such that the functional and safety performance requirements are satisfied.
5. Appropriate documentation of the integration testing of the SRECS should be produced, stating the test results and whether the objectives and criteria specified during the design and development phase have been met. If there is a failure, the reasons for the failure should be documented, corrective action taken and re-testing carried out. During the integration and testing, any modification or change to the SRECS should be subject to an impact analysis that should identify all components affected and additional verification.

Regardless of the SRECS sub-system design / development approach taken, the SRECS should be installed implemented in accordance with the requirements of the final system validation (AS(IEC)61508 Phase 13). Appropriate records of the installation of the SRECS should be produced, stating any test results. If there is a failure, the reasons for the failure should be recorded.

Documentation should be provided by the designer for inclusion in the Safety Lifecycle Dossier, including:

1. Results of SRECS Sub-system Hardware Testing,
2. Documentation of SRECS Hardware Integration Testing,
3. Results of Application Software Module Testing,
4. Results of Application Software Integration Testing,
5. Documentation of SRECS Installation and Commissioning,

7.13 AS(IEC)61508 Phase 10 - E/E/PE Safety-related Systems Realisation (using ISO13849)

Note: See Section 4.4 regarding the validity of the design and development approach provided in Part 1501 and 1502 of this standard. It is my view that the approach to the specification and realisation of machinery-sector safety-related systems provided by the latest issues of ISO13849 or AS(IEC)62061 should be used, and the out-dated approaches provided AS4024.1501-2006 and AS4024.1502-2006 should be avoided where possible. AS4024.1 was in the process of being updated by Standards Australia at the time of writing of this book. It is anticipated that AS4024.1 may be re-issued in 2013 or 2014 taking account of the latest issues of ISO12100, ISO14121 and ISO13849.

Objective/s:

Alternatively, the design and development of the E/E/PE safety-related systems should follow a seven (7) step process, as provided by ISO13849-1 Figure 3, shown below in Figure 7.14.

Guidance on the implementation of this process is provided in ISO-13849-1 Clauses 4.3 to 4.7.

Figure 7.14: ISO13849 Design and Development Process

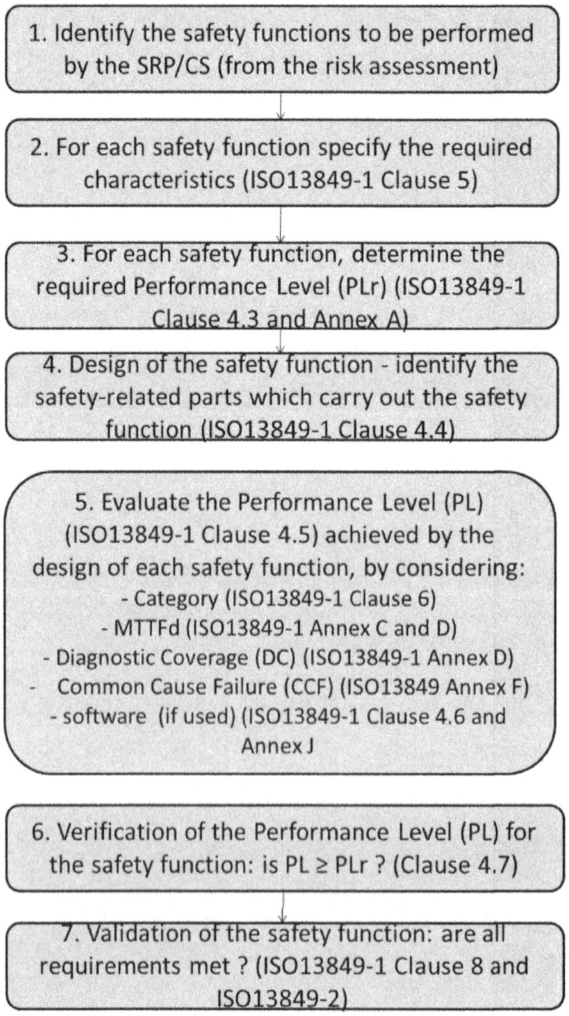

Responsible Entity:

Whilst ultimate responsibility for ensuring completion of the requirements of these phases rests with the end-user (eg. the mining company), responsibility for conducting the activities may be delegated to the vendor or designer.

Activities to be Undertaken / Documentation to be Kept:

ISO13849-1 Step 1 – Identify the Safety Functions
ISO13849-1 Step 2 –Specify the Required Characteristics
ISO13849-1 Step 3 – Determine the Required Performance Level (PL_r)

Steps 1, 2 and 3 of the ISO13849-1 SRP/CS design and development process is achieved by the drafting of the Safety Requirements Specification (SRS) where the functional intent, inputs and outputs, intended realisation and operating parameters of each SRCF are documented along with the required PL.

ISO13849-1 Step 4 – Design of the Safety Function

Step 4 comprises the 'high-level' architectural 'block diagram' design of the Safety Functions. Step 4 of ISO13849-1 is equivalent to Steps 2 to 4 of AS(IEC)62061 (see Figure 7.6). ISO13849-1 Annex B provides guidance for the development of the block diagrams.

In ISO13849-1 Safety Functions are normally viewed as being implemented by a number of SRP/CS's that are grouped into input, logic and output sub-systems in the same manner as AS(IEC)61508 / AS(IEC)62061 (see Figure 7.6). The basic concept of the block diagram is shown in Figure 7.15 below, however in some cases the input, logic and output sub-systems may be provided by the same SRP/CS.

Figure 7.15: ISO13849-1 Block Diagram

Regardless of the architecture of the safety function, the most important part of this step is to identify which SRP/CS's are included in each safety function. Generally, if a fault in the SRP/CS will lead to a failure of the safety function then the SRP/CS shall be included as part of the safety function.

The outputs of Step 4 are 'high-level' 'block diagrams' which describe the architecture, as in AS(IEC)61508 / AS(IEC)62061. For simple Safety Functions this may comprise three (3) functional blocks in series, but it is often more complex than this (eg. a redundant sub-system architecture using two different input SRP/CS's – a limit switch and a proximity switch).

A 'SRP/CS Concept Design' document should be initiated with the following information, and a copy should be provided by the designer for inclusion in the Safety Lifecycle Dossier.

1. A graphical representation of the Safety Functions and SRP/CS 'block diagram' architecture.
2. A definition of the inputs and outputs of each SRP/CS block.

3. The safety requirements (ie. functional and safety integrity) allocations for each SRP/CS block.
4. Identification of those SRP/CS blocks that will involve a complement of software.
5. A SRP/CS Safety Requirements Specification. If the principal SRP/CS designer intends to delegate or sub-contract the design of any SRPS/CS to another organisation, they should formally communicate the SRP/CS safety requirements to the sub-contractor / delegate in a "SRP/CS Safety Requirements Specification".

ISO13849-1 Step 5 – Evaluate the Performance Level (PL) Achieved

Step 5 comprises an evaluation of the 'high-level' architectural design against the requirements of the allocated PL. It is normal for Steps 4 and 5 to be conducted iteratively. Steps 5 and 6 of ISO13849-1 are equivalent to Step 8 of AS(IEC)62061 (see Figure 7.6).

Issues of ISO13849-1 since 2006 have used PL as the safety integrity target measure.

Four (4) parameters are used to verify the PL achieved by a safety function:

1. **Architecture**, determined according to a set of designated CAT B, 1, 2, 3, and 4 architectures, provided in Clause 6.
2. **Mean Time to Dangerous Failure ($MTTF_d$)** of a "single channel"[43], determined according to the calculation methodology provided in Annexes C and D
3. **Average Diagnostic Coverage (DC_{ave}),** determined according to table provided in Annex E
4. **Common Cause Failure (CCF) score**, determined according to a table provided in Annex F

The referenced parts of ISO13849-1 provide guidance on the methods to be used to determine these parameters.

The achieved PL is then determined via the chart at ISO13849-1 Figure 5 (shown here in Figure 7.16), which combines the above four (4) parameters. The chart is intended to be read 'bottom-up'. For example, a safety function having CAT3 architecture, CCF score $\geq$ 65, $DC_{ave} \geq 60\%$ and $MTTF_d \geq 30$ years would be deemed capable of PLd performance. In Figure 7.16 there are actually six (6) combinations of parameters that would allow PLd to be achieved. Designers usually have several options available to them when designing a safety function to a PL target.

[43] In the ISO13849-1 concept of 'designated architectures' each series of Input-Logic-Output SRP/CS's is a "channel". Under this concept the designated architectures for CATB, 1 and 2 are single channel architectures and those for CAT3 and 4 are dual-channel architectures. For the dual-channel (or other redundant architectures) ISO13849-1 requires that the $MTTF_d$ for each channel be determined and the lowest $MTTF_d$ for the channels be selected as the $MTTF_d$ parameter.

Figure 7.16: PL Verification

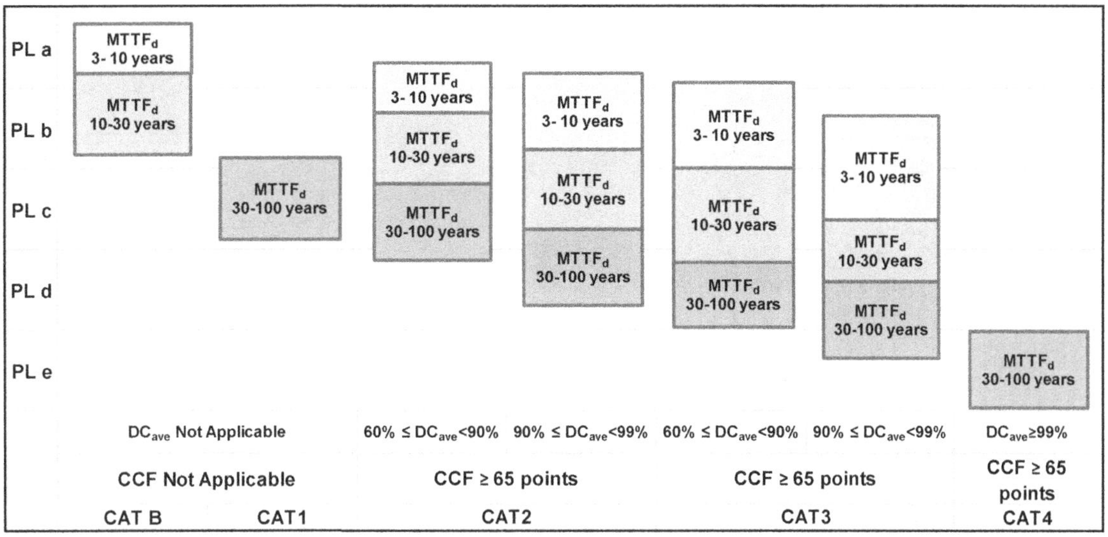

However, if you find this chart to be ambiguous or confusing, you are not alone.

For example a safety function having CAT3 architecture, a CCF score ≥ 65, a $DC_{ave} \geq 90\%$ and an $MTTF_d \geq 30$ years denotes a level of safety integrity in either the PLd or PLe regions. No further guidance is provided in this table to allow further discrimination of the safety function as either PLd or PLe. In this circumstance the final discrimination should be based on ISO13849-1 Annex K, Table K.1. This table suggests that if the PFH is less than 10^{-7} PLe is achieved and if the PFH is greater than 10^{-7} PLd is achieved.

There are subtle complexities and a maze of rules involved in PL evaluation, particularly for complex safety functions. I <u>strongly recommend</u> that *SISTEMA* software should be used during the iterative design and evaluation process of Steps 4 and 5. *SISTEMA* provides comprehensive support for evaluating, documenting and reporting on safety-related systems against the requirements of ISO13849-1 for PLa to PLe. *SISTEMA* is particularly useful for analysing safety functions with multiple channel architectures or many SRP/CS's.

SISTEMA software is from the *Institut fuer Arbeitsschutz der Deutschen Gesetzlichen Unfallversicherung (IFA)* in Germany. IFA is an institute for research and testing within German Social Accident Insurance. *SISTEMA* software may be downloaded free of charge from:

http://www.dguv.de/ifa/en/pra/softwa/sistema/index.jsp

Component libraries have been developed by several safety component manufacturers for use with *SISTEMA*. A list of manufacturer's component libraries can be accessed via http://www.dguv.de/ifa/en/pra/softwa/sistema/bibliotheken/index.jsp

A screenshot of the *SISTEMA* software graphical user interface is provided in Figure 7.17 to illustrate its PL evaluation capabilities.

Figure 7.17: SISTEMA Software Tool

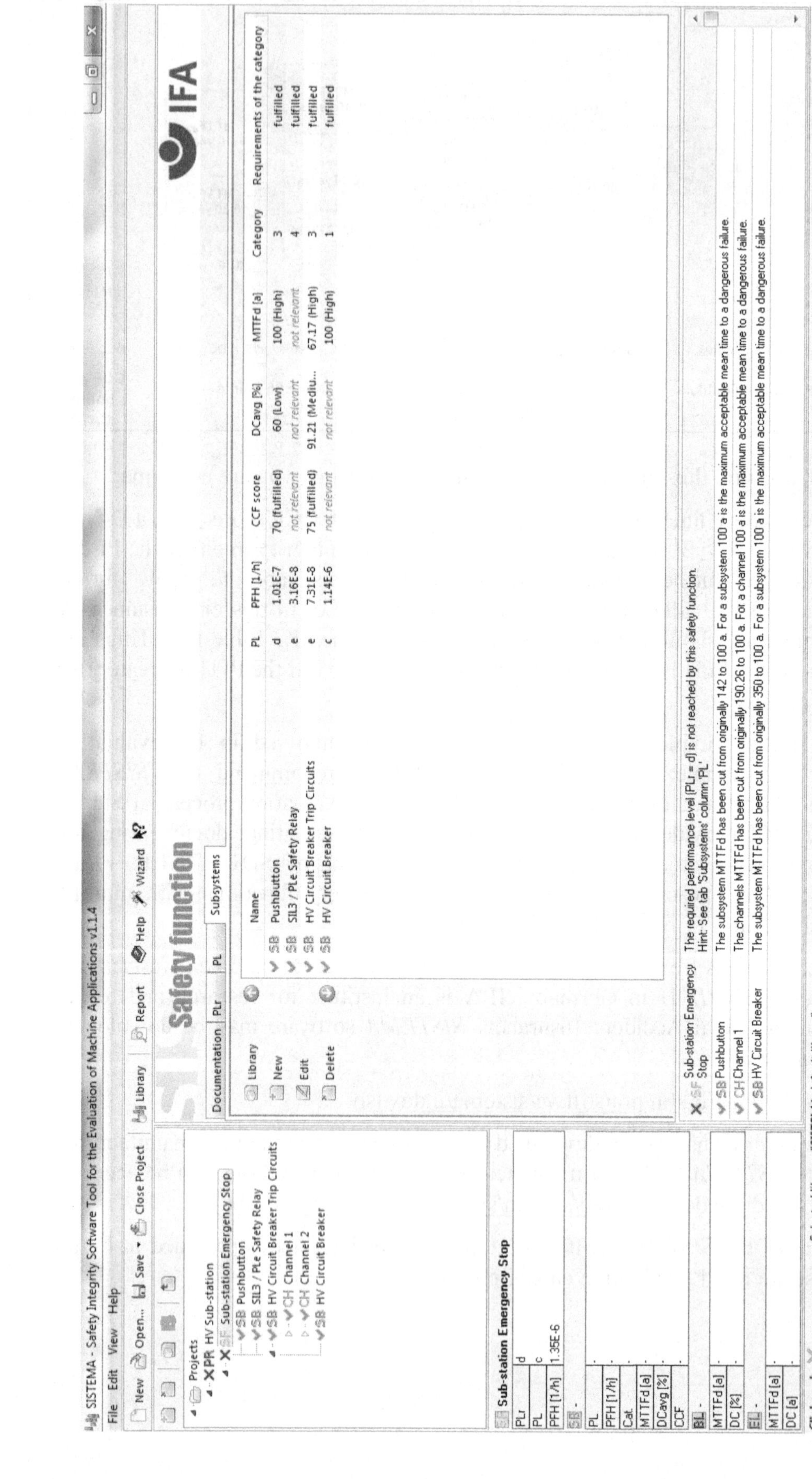

A special circumstance exists where the PL's of all of the SRP/C'S's have been previously verified (eg. the designer has used all previously PL-certified components in the design). In this case, the achieved PL of the overall safety function can be determined by the use of ISO13849-1 Table 11 alone (shown here in Table 7.6). No other analysis (ie. $MTTF_d$, CCF, DC_{ave}) is required. However this method assumes that the safety function has a simple series architecture (as in Figure 7.15).

Table 7.6: Determining PL via ISO13849-1 Table 11

PL_{low} (lowest PL achieved by the component SRP/CS's)	N_{low} (number of SRP/CS's with the lowest PL)	Overall PL
a	>3	None
	≤3	a
b	>2	a
	≤2	b
c	>2	b
	≤2	c
d	>3	c
	≤3	d
e	>3	d
	≤3	e

To use this table, the PL of each SRP/CS used by the safety function are obtained. Then the following are evaluated:

- The lowest PL achieved by all of the SRP/CS's
- The number of SRP/CS's that achieve this PL.

Combining these two parameters in the table allows an overall achieved PL to be determined for the safety function. An example of the use of this table is shown in Figure 7.18.

Figure 7.18: Example of Determining PL via ISO13849-1 Table 11

In this example PL_{low} = PLc and $N_{low} = 2$, so the overall PL achieved by the safety function according to ISO13849-1 Table 11 is PLc. The overall PFH of the safety function is determined by adding the individual PFH's of the SRP/CS's.

A designer may find that certain SRP/CS's have been previously certified to a SIL instead of a PL. In this case, the SIL may be converted to the equivalent PL using ISO13849-1 Table 4 (shown in Table 5.2). Therefore, if the designer has used all previously SIL-certified components in the design, these can be converted to PL's and ISO13849-1 Table 11 used to determine the overall PL.

In determining the achieved PL the following aspects of the design of each safety function must also be considered:

- Systematic failure avoidance and control, per Annex G
- Safety-related software, per Clause 4.6 and Annex J

A summary of the ISO13849-1 requirements for systematic failure avoidance and control and software design and development are provided in Sections 7.15 and 7.16.

Furthermore, the designer must also consider the ability of the SRP/CS's to perform the safety function under the expected environmental conditions.

There is no specific guidance in ISO13849-1 regarding the equivalence of PL's and SIL's for low demand safety functions.

However, I suggest that a common-sense should be applied, as follows:

- It is reasonable to expect that mining and industrial machinery will require a mix of low demand, high demand and continuous mode safety functions
- The PL / SIL equivalences stated in the latest issue of ISO13849-1 should be taken to be valid for high and continuous demand mode safety functions
- The PL / SIL equivalences stated in the latest issue of ISO13849-1 may be taken to be valid for low demand mode safety functions, on the condition that a proof-testing regime is established which would allow the achievement of a PFD in the same SIL range.

The Safety Lifecycle Dossier should be updated with the following information:

1. A PL Evaluation Report addressing the aforementioned features of each safety function.
2. A statement of the analyst's competence.

ISO13849-1 Step 6 – Verification

Verification comprises a confirmation that the required PL (PL_r) for each safety function has been achieved by the design. Verification should be performed by persons with an appropriate degree of independence to those who performed the design work. It is suggested that the degree of independence of the design verifier be commensurate with the guidelines provided in AS(IEC)61508.1 (see Section 6.5) for the SIL that is equivalent to the required PL (eg. for PLd apply the independence requirements of SIL2).

Step 6 is significantly assisted if *SISTEMA* software has been used during Step 5. The *SISTEMA* file can be reviewed by the verifier for accuracy and completeness.

The Safety Lifecycle Dossier should be updated with the following information:

1. Verification Report. This should comprise an audit checklist based on the objectives and requirements of each preceding step, including observations made and corrective actions required.
2. A statement of the verifier's competence and independence.

ISO13849-1 Step 7 – Validation

The requirements for validation are provided in ISO13849-1 Clause 8 and ISO13849-2. These provide detailed requirements for the planning and conduct of the validation activities for the safety-related systems and their environmental and maintenance requirements.

There is a transition from ISO13849 requirements to AS(IEC)61508 requirements during this phase. If ISO13849-1 is being used within the framework of an overall AS(IEC)61508 functional safety lifecycle, then the option may exist to conduct the validation step according to that standard (see Section 7.21).

The Safety Lifecycle Dossier should be updated with the following information:

1. Identification of the version of the SRS under which the validation was carried out
2. Identification of the safety functions being validated (by test or analysis)
3. Tools and equipment used with calibration data
4. Documentation of the validation activities
5. The results of the validation activities, including recorded discrepancies between expected and actual results
6. Configuration identification of the item under test, the procedures applied and the test environment
7. A statement of the validator's competence and independence

7.14 AS(IEC)61508 Phase 10 – Data Communications

The 2010/2011 issue of AS(IEC)61508.2 imposes additional requirements on designers of safety–related systems that incorporate data communications.

Most of the requirements of the previous issue remain in place, however AS(IEC)61508.2 Clause 7.4.11 now provides two design concepts for the architecture of data communications systems and networks used within a safety-related system.

Specifically, this clause has introduced the concepts of "white channel" and "black channel" data communications. One of these approaches must be taken in order for the data communications process of a safety function to satisfy AS(IEC)61508.

"White" Channel

"White channel" data communications where all parts of the communication channel are designed and validated according to AS(IEC)61508.

The entire communication channel including the network components (eg. network switches, access points, wireless transmitters / receivers etc…), protocols and services must be designed, implemented and validated according to AS(IEC)61508 and either:

- IEC61784-3: Industrial communication networks – Profiles – Functional safety fieldbuses, or
- IEC62280: Railway applications - Communication, signalling and processing systems - Safety-related communication in closed / open transmission

This concept is shown in Figure 7.19 below.

Figure 7.19: "White" Channel Data Communications

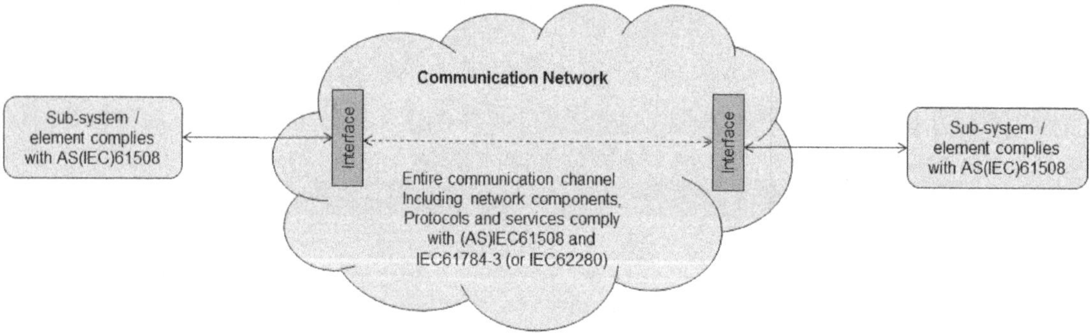

IEC61784-3 strongly recommends that the each individual communications link used by a safety function does not consume more than 1% of the required PFD/PFH of the safety function. For example, this means that each communication link used by a SIL2 safety function cannot have a PFD of more than 10^{-4} or a PFH of more than 10^{-8}. If this cannot be achieved with single communications links, then redundant links may be needed.

Providing a redundant safety network may provide technical and financial challenges for larger, more complex safety networks. Redundant communications may be achieved by the use of a "ring" network topology, as shown in Figure 7.20. Two (2) communication paths are provided for communications between any two (2) interface devices and using considerably less network hardware than would be required with an equivalent "star" network topology.

Having 2 redundant communications paths would significantly decrease the overall PFD or PFH of the communications process. However, it would also mean that the rate of spurious failure (or safe failure) of the communications process between two points in the network would effectively double. Therefore, spurious operation of a safety function due to communication failure would effectively double also.

Figure 7.20: Ring Network Topology for Redundant Data Communications

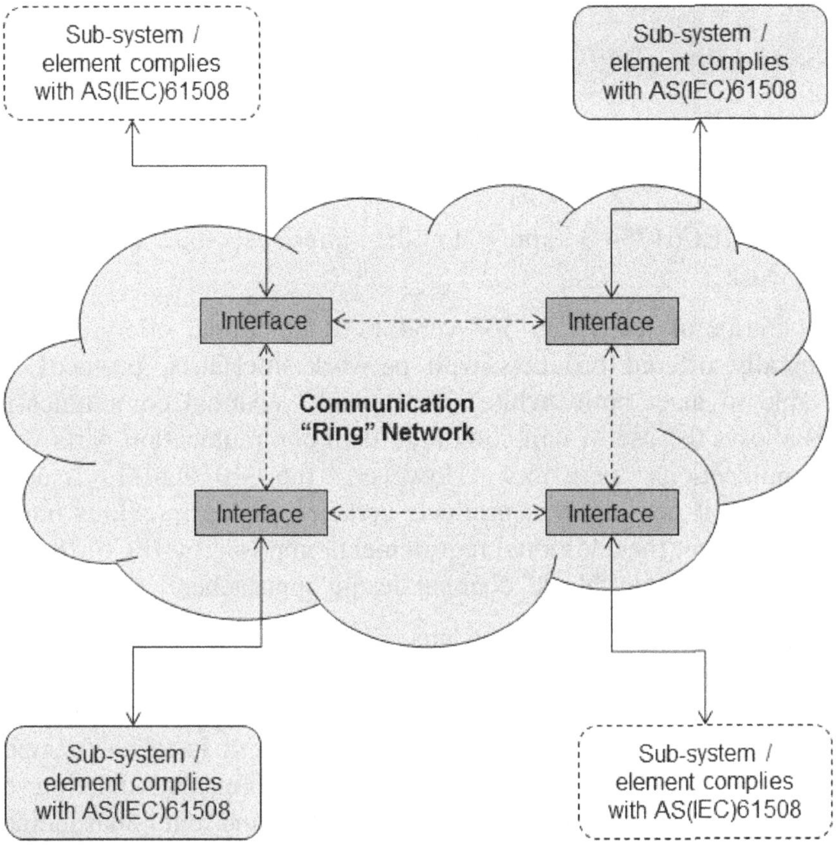

"Black" Channel

The "black" channel approach applies where parts of the communication channel such as network switches, access points, wireless transmitters / receivers etc…are not designed or validated according to AS(IEC)61508.

With "black" channel communications the data from the sending element is transmitted into an unknown communications channel. Therefore, the data and the network elements must have built-in mechanisms to detect errors or interference with a confidence that is suitable for safety applications. This might be achieved by additional diagnostics or by application functions such as 'handshaking' routines.

In this approach, the use of non-compliant devices within the communication channel may occur, however the interfaces between the sub-systems and elements that interface with the communication channel must be designed in accordance with either:

- IEC61784-3: Industrial communication networks – Profiles – Functional safety fieldbuses, or
- IEC62280: Railway applications - Communication, signalling and processing systems - Safety-related communication in closed / open transmission

This is shown in Figure 7.21 below.

Figure 7.21: "Black Channel" Data Communications

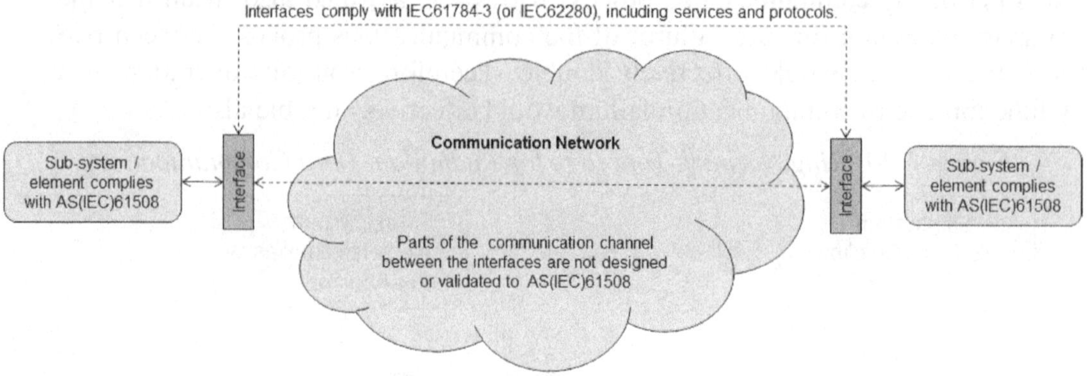

The requirements of IEC61784-3 apply to the interfaces but not the data communications network.

Since the 2010/2011 issue of AS(IEC)61508 vendors of functional safety compliant products have typically offered products with network interfaces, protocols and services that are able to meet both "white" and "black" channel communications requirements. This allows the use of non-compliant data communication parts within safety data communications networks. However, the 2010/2011 issue of AS(IEC)61508 has made it possible that products certified via the previous issue of AS(IEC)61508 may not meet the additional requirements imposed by IEC61784-3 (or IEC62280) for the "white" and/or "black" channel design approaches.

Regardless of the design approach taken, Clause 7.4.11 requires that the "failure measure" of the communications process (eg. the residual error rate) must be estimated and taken into account when estimating the probability of failure of the overall safety function. Failure of the communications process is taken to include transmission errors and processing errors such as data repetition, data deletion, data insertion, data re-sequencing, data corruption, data delay and data masquerade[44] errors.

[44] Masquerade errors are those in which the true source of a message is not correctly identified.

7.15 AS(IEC)61508 Phase 10 - Hardware Systematic Failure Avoidance and Control

In AS61508.4 "systematic failure" has been defined as:

"failure, related in a deterministic way to a certain cause, which can only be eliminated by a modification of the design or of the manufacturing process, operational procedures, documentation or other relevant factors".

Each of the standards defines a number of measures and techniques that may be used to avoid systematic failures or control their effect at each phase of the functional safety lifecycle. These techniques and measures are broadly divided by relevance to hardware or software systematic failure.

Demonstration of compliance involves the gathering of evidence of an appropriate application of the specified activities during the various phases of the safety lifecycle. This allows a deterministic argument to be made that the requirements for systematic failure avoidance and control at the target SIL or PL have been met.

This section summarises the approaches to hardware systematic failure avoidance and control in AS(IEC)61508.2, AS(IEC)62061 and ISO13849. AS(IEC)61511.2 provides similar measures to AS(IEC)61508, but these are embedded in the text of AS(IEC)61511.1 rather than in a separate annex. Systematic failure avoidance and control of safety-related software is discussed in the next section.

AS(IEC)61508.2

AS(IEC)61508.2 Annexes A and B detail the requirements for avoidance and control of systematic failures for hardware used in E/E/PE safety-related systems.

Table 7.7 provides details on the techniques and measures for hardware systematic failure avoidance and control that are mandatory or highly recommended by AS(IEC)61508.2. The purpose of this table is to draw the reader's attention to those techniques and measures that are the most critical to compliance with AS(IEC)61508.2 for SIL1, SIL2 and SIL3. A range of other techniques and measures are recommended by AS(IEC)61508.2, but are arguably less critical than these. Designers should consult the standard and select techniques and measures, as appropriate to the application.

Table 7.7: AS(IEC)61508 'M' and 'HR' Techniques & Measures (Hardware)

Technique / Measure	AS(IEC)61508-7 Reference	Interpretation of the Requirement	SIL1	SIL2	SIL3
Control of Systematic Failures Caused by Hardware and Software Design					
Program Sequence Monitoring	A.9	Temporal and logical monitoring of the program sequence at many checking points of the program sequence.	Yes	Yes	Yes
Control of Systematic Failures Caused by Environmental Stresses or Influences					
Measures against voltage breakdown, voltage variation, over-voltage and low voltage	A.8	Over-voltage protection, secondary voltage monitoring, power-down with safety shut-off.	Yes	Yes	Yes

Technique / Measure	AS(IEC)61508-7 Reference	Interpretation of the Requirement	SIL1	SIL2	SIL3
Separation of electrical energy lines from information lines	A.11.1	Minimise cross-talk induced by high currents in the information lines.	Yes	Yes	Yes
Increase of interference immunity	A.11.3	Filtering and shielding to achieve immunity levels specified in AS(IEC)62061 Annex E and IEC61000.	Yes	Yes	Yes
Measures against the physical environment (eg. temperature, humidity, dust, water, vibration, corrosives)	A.14	Ingress Protection (IP) ratings against the expected environment.	Yes	Yes	Yes
Program Sequence Monitoring	A.9	Temporal and logical monitoring of the program sequence at many checking points of the program sequence.	Yes	Yes	Yes
Measures against temperature increase	A.10	Temperature sensing, fan control, actuation of the safety shut-off via a thermal fuse, condition alarms, forced air cooling.	Yes	Yes	Yes
Spatial separation of multiple lines	A.11.2	Minimise cross-talk induced by high currents in multiple lines.	Yes	Yes	Yes
Control of Systematic Operational Failures					
Modification protection	B.4.8	Plausibility checks on sensor signals; detection via the technical process; automatic start-up tests.	Yes	Yes	Yes
Avoidance of Mistakes During Specification					
Project Management	B.1.1	Validation independent of design; project monitoring; standardised validation procedure; configuration management; failure statistics; computer aided engineering; computer-aided software engineering.	Yes	Yes	Yes
Documentation	B.1.2	Guidelines for consistent content and layout across organisation; contents checklists; computer-aided documentation management; formal change control.	Yes	Yes	Yes
Separation of E/E/PE safety-related systems from non safety-related systems	B.1.3	Total separation, ie. no write access from non safety-related systems to E/E/PE safety-related systems and separate physical locations to avoid common-cause influences.	Yes	Yes	Yes

Technique / Measure	AS(IEC)61508-7 Reference	Interpretation of the Requirement	SIL1	SIL2	SIL3
Structured specification	B.2.1	Hierarchical separation described using computer-aided engineering tools; automatic consistency checks; refinement down to functional level.	Yes	Yes	Yes
Inspection of the specification	B.2.6	Inspection and re-inspection by an independent organisation using a formal procedure with correction of all faults found.		Yes	Yes
Semi-formal methods	B.2.3, see also Table B.7 of AS(IEC)61508.3	Describing E/E/PE safety-related systems with different semi-formal methods to show different aspects; consistency check between the methods.			Yes
Avoidance of Introducing Faults During Design and Development					
Observance of guidelines and standards	B.3.1	Standards and guidelines which are specific for the application and sector (eg. AS4240 – Remote Controls for Mining Equipment, AS1755 – Conveyors – Safety Requirements, IECEx requirements etc...).	Yes	Yes	Yes
Project Management	B.1.1	As above.	Yes	Yes	Yes
Documentation	B.1.2	As above.	Yes	Yes	Yes
Structured Design	B.3.2	Re-use of tested circuit parts; traceability between specification, design, circuit diagram, and parts lists; computer-aided; based on defined methods.	Yes	Yes	Yes
Modularisation	B.3.4	Re-use of well proven modules, easily comprehensible modules; each module has a maximum of one input, one output and one failure exit.	Yes	Yes	Yes
Semi-formal methods	B.2.3, see also Table B.7 of AS(IEC)61508.3	Describing E/E/PE safety-related systems with different semi-formal methods to show different aspects; consistency check between the methods.			Yes
Avoidance of Faults During Integration					
Functional Testing	B.5.1	Inputs per normally expected operation, outputs observed and compared with the specification. Deviations documented.	Yes	Yes	Yes
Project Management	B.1.1	As above.	Yes	Yes	Yes
Documentation	B.1.2	As above.	Yes	Yes	Yes

Technique / Measure	AS(IEC)61508-7 Reference	Interpretation of the Requirement	SIL1	SIL2	SIL3
Avoidance of Faults and Failures During Operation and Maintenance Procedures					
Operation and Maintenance Instructions	B.4.1	Easily understood user instructions. Figures and schematics for complex procedures.	Yes	Yes	Yes
User Friendliness	B.4.2	Reduced complexity for users.	Yes	Yes	Yes
Maintenance Friendliness	B.4.3	Reduced complexity for maintainers.	Yes	Yes	Yes
Project Management	B.1.1	As above.	Yes	Yes	Yes
Documentation	B.1.2	As above.	Yes	Yes	Yes
Limited Operation Possibilities	B.4.4	Defined, robust procedure for allowing operation.			Yes
Protection against operator mistakes	B.4.6	Confirmation and consistency checks on each channel.			Yes
Avoidance of Faults During E/E/PE Safety-related System Safety Validation					
Functional testing	B.5.1	Inputs per normally expected operation, outputs observed and compared with the specification. Deviations documented.	Yes	Yes	Yes
Functional testing under environmental conditions	B.6.1	Functional testing under conditions equivalent to IP / EMI ratings.	Yes	Yes	Yes
Interference surge immunity testing	B.6.2	Surge immunity shall be demonstrably higher than the boundary values for real operating conditions.	Yes	Yes	Yes
Fault insertion testing when required DC ≥ 90%	B.6.10	Simulate faults and document the response. At component level, including boundary data.	Yes	Yes	Yes
Project Management	B.1.1	As above.	Yes	Yes	Yes
Documentation	B.1.2	As above.	Yes	Yes	Yes
Expanded functional testing	B.6.8	Tests that all safety functions are maintained in the case of static input states and/or unusual input changes, caused by faulty process or operating conditions (including those that may be very rare).		Yes	Yes

AS(IEC)62061

AS(IEC)62061 Clauses 6.4 and 6.7.9 detail the requirements for avoidance and control of systematic failures for hardware used in E/E/PE safety-related systems and their sub-systems. The following is a summary of the requirements:

Requirements for the Avoidance of Systematic Hardware Failures of the SRECS

1. Design and implement in accordance with the Functional Safety Management Plan.
2. Proper selection, combination, arrangement, assembly and installation of sub-systems including cabling, wiring and inter-connections.
3. Use within the manufacturer's specifications.
4. Use of manufacturer's application notes (eg. catalogue sheets, installation instructions) and use of good engineering practice.
5. Use of sub-systems that have compatible operating characteristics.
6. Electrical protection in accordance with AS60204-1.
7. Prevention of the loss of functional earth connection in accordance with AS60204-1.
8. Undocumented modes of component operation shall not be used (eg. unreserved registers of programmable equipment).
9. Consideration of foreseeable misuse, environmental changes or modifications.
10. At least one of the following:
    a. Hardware design review (by inspection or walk-through).
    b. Use of advisory tools or computer-aided design packages capable of simulation and analysis, and/or the use of computer-aided design tools to perform the design procedures systematically with the use of pre-designed elements that are already available and tested.
    c. Systematic and complete simulation of a SRECS design in terms of both functional performance and correct dimensioning and interaction of its sub-systems.

Requirements for Avoidance of Systematic Failures of SRECS Sub-systems

1. Proper selection, combination, arrangement, assembly and installation of components including cabling, wiring and inter-connections. Apply manufacturer's application notes (eg. catalogue sheets, installation instructions) and use of good engineering practice.
2. Use of the sub-system and the sub-system elements within the manufacturer's specification and installation instructions.
3. Use components with compatible operating characteristics.
4. Withstanding specified environmental conditions – design the sub-system so that it is capable of working in all expected environments and in any foreseeable adverse conditions (eg. temperature, humidity, vibration, electro-magnetic interference).

5. Use components that are in accordance with an appropriate standard and have their failure modes well-defined – to reduce the risk of undetected faults by the use of components with specific characteristics.
6. Use of suitable materials and adequate manufacturing – selection of material, manufacturing methods, and treatment in relation to eg. stress, durability, elasticity, friction, wear, corrosion, temperature, conductivity, di-electric strength.
7. Correct dimensioning and shaping – consider the effects of eg. stress, strain, fatigue, temperature, surface roughness, manufacturing tolerances.
8. At least one of the following:
    a. Hardware design review (by inspection or walk-through).
    b. Use of computer-aided design tools capable of simulation and analysis. Perform the design procedures systematically and include appropriate automatic construction elements that are already available and tested.
    c. Systematic simulation of a sub-system design in terms of both functional performance and correct dimensioning of their components.

Requirements for the Control of Systematic Hardware Failures of the SRECS

1. Use of de-energisation principle, ie. the safe state is achieved with loss of electrical supply.
2. Measures to control the effect (ie. does not lead to a hazard) of temporary sub-system failures, eg.
    a. Voltage variations (ie. interruptions, dips).
    b. Electromagnetic interference.
3. Measures to control the effects of errors and other effects arising from any data communication process, including transmission errors, repetitions, deletion, insertion, re-sequencing, corruption, delay and masquerade (true content of a message is not correctly identified).
4. When a dangerous fault occurs at an interface, the fault reaction function shall be performed before the hazard due to this fault can occur. When a fault that reduces hardware fault tolerance to zero occurs, this fault reaction shall take place before the estimated Mean Time To Repair (MTTR) is exceeded.

Requirements for the Control of Systematic Failures of SRECS Sub-systems

1. Measures to control the effects of insulation breakdown, voltage variations and interruptions, over-voltage and under-voltage - sub-system behaviour in response to these shall be pre-determined so that the sub-system can achieve or maintain a safe state of the SRECS.
2. Measures to control or avoid the effects of the physical environment (eg. temperature, humidity, water, vibration, dust, corrosive substances, electro-magnetic interference and its effects) - sub-system behaviour in response to these shall be pre-determined so that the sub-system can achieve or maintain a safe state of the SRECS.

3. Measures to control or avoid the effects of temperature increase or decrease if the variations can occur – the sub-system should be designed so that temperature variations can be detected before it begins to operate outside specifications.
4. At least one of the following:
    a. Fault detection by on-line monitoring,
    b. Tests by comparison of redundant hardware,
    c. Diverse hardware (eg. a proximity switch and a limit switch)
    d. Operation in the positive mode (eg. limit switch pushed as the guard is opened),
    e. Oriented mode of failure (pre-dominant failure mode is known in advance and the effect is always the same),
    f. Over-dimensioning by a suitable factor, where the manufacturer can demonstrate that de-rating will improve reliability. Where this measure is used, an over-dimensioning factor of at least 1.5 should be used.

Electromagnetic Immunity

1. Conformance to the requirements of IEC61000-6-2 for the electro-magnetic phenomena given in AS62061 Annex E, provided that:
    a. Unsafe conditions or hazards are not introduced.
    b. There is no loss of the Safety-related Control Functions (SRCF's).
    c. The SRCF's implemented by the SRECS may be disturbed temporarily or permanently provided that a safe state of the machine is maintained or achieved before a hazard can occur.
    d. Where electro-magnetic phenomena can result in the destruction of components, it shall be ensured (eg. by analysis) that functional safety is not affected, including by lower values of electro-magnetic phenomena that can cause partial destruction.

ISO13849

ISO13849-1 Annex G details the requirements for avoidance and control of systematic failures for hardware used in SRP/CS. The following is a summary of the requirements:

Requirements for Avoidance of Systematic Hardware Failures

The following measures should be applied.

1. Use of suitable materials and adequate manufacturing (selection of material, manufacturing methods and treatment in relation to, e.g. stress, durability, elasticity, friction, wear, corrosion, temperature, conductivity, dielectric rigidity).
2. Correct dimensioning and shaping (consideration of stress, strain, fatigue, temperature, surface roughness, tolerances, manufacturing).
3. Proper selection, combination, arrangements, assembly and installation of components, including cabling, wiring and any interconnections (apply appropriate standards and manufacturer's application notes, e.g. catalogue sheets, installation instructions, specifications, and use of good engineering practice).
4. Compatibility (use components with compatible operating characteristics).
5. Withstanding specified environmental conditions (design the SRP/CS so that it is capable of working in all expected environments and in any foreseeable adverse conditions, e.g. temperature, humidity, vibration and electromagnetic interference (EMI).
6. Use of components designed to an appropriate standard and having well-defined failure modes (to reduce the risk of undetected faults by the use of components with specific characteristics

In addition, one or more of the following measures should be applied, taking into account the complexity of the SRP/CS and its PL.

1. Hardware design review (e.g. by inspection or walk-through - to reveal by reviews and analysis discrepancies between the specification and implementation
2. Computer-aided design tools capable of simulation or analysis (perform the design procedure systematically and include appropriate automatic construction elements that are already available and tested.
3. Simulation (perform a systematic and complete inspection of an SRP/CS design in terms of both the functional performance and the correct dimensioning of their components).

Requirements for Control of Systematic Hardware Failures

The following measures should be applied.

1. Use of de-energization (the safety-related parts of the control system (SRP/CS) should be designed so that with loss of its power supply a safe state of the machine can be achieved or maintained).
2. Measures for controlling the effects of voltage breakdown, voltage variations, overvoltage, undervoltage (SRP/CS behaviour in response to voltage breakdown, voltage variations, overvoltage, and undervoltage conditions should be

predetermined so that the SRP/CS can achieve or maintain a safe state of the machine).

3. Measures for controlling or avoiding the effects of the physical environment (for example, temperature, humidity, water, vibration, dust, corrosive substances, electromagnetic interference and its effects - SRP/CS behaviour in response to the effects of the physical environment should be predetermined so that the SRP/CS can achieve or maintain a safe state of the machine).
4. Program sequence monitoring shall be used with SRP/CS containing software in order detect defective program sequences (a defective program sequence exists if the individual elements of a program (e.g. software modules, subprograms or commands) are processed in the wrong sequence or period of time or if the clock of the processor is faulty.
5. Measures for controlling the effects of errors and other effects arising from any data communication process.

In addition, one or more of the following measures should be applied, taking into account the complexity of the SRP/CS and its PL:

1. Failure detection by automatic tests;
2. Tests by redundant hardware;
3. Diverse hardware;
4. Operation in the positive mode;
5. Mechanically linked contacts;
6. Direct opening action;
7. Oriented mode of failure;
8. Over-dimensioning by a suitable factor, where the manufacturer can demonstrate that de-rating will improve reliability (where over-dimensioning is appropriate, an over-dimensioning factor of at least 1.5 should be used).

Requirements for Avoidance of Systematic Failures During Integration of the SRP/CS

The following measures should be applied during integration of the SRP/CS:

1. Functional testing.
2. Project management.
3. Documentation.
4. Black-box testing, taking into account the complexity of the SRP/CS and its PL.

7.16 AS(IEC)61508 Phase 10 – Software Systematic Failure Avoidance and Control

This section summarises the approaches to software systematic failure avoidance and control in AS(IEC)61508.3, AS(IEC)62061 and ISO13849. AS(IEC)61511.2 provides similar measures to AS(IEC)61508, but these are embedded in the text of AS(IEC)61511.1 rather than in a separate annex. The common element in all of these approaches is the use of the software "V-Model" for design and development, verification and validation, as shown in Figure 7.22 (from AS(IEC)61508.3 Figure 6).

Figure 7.22: Software "V-Model"

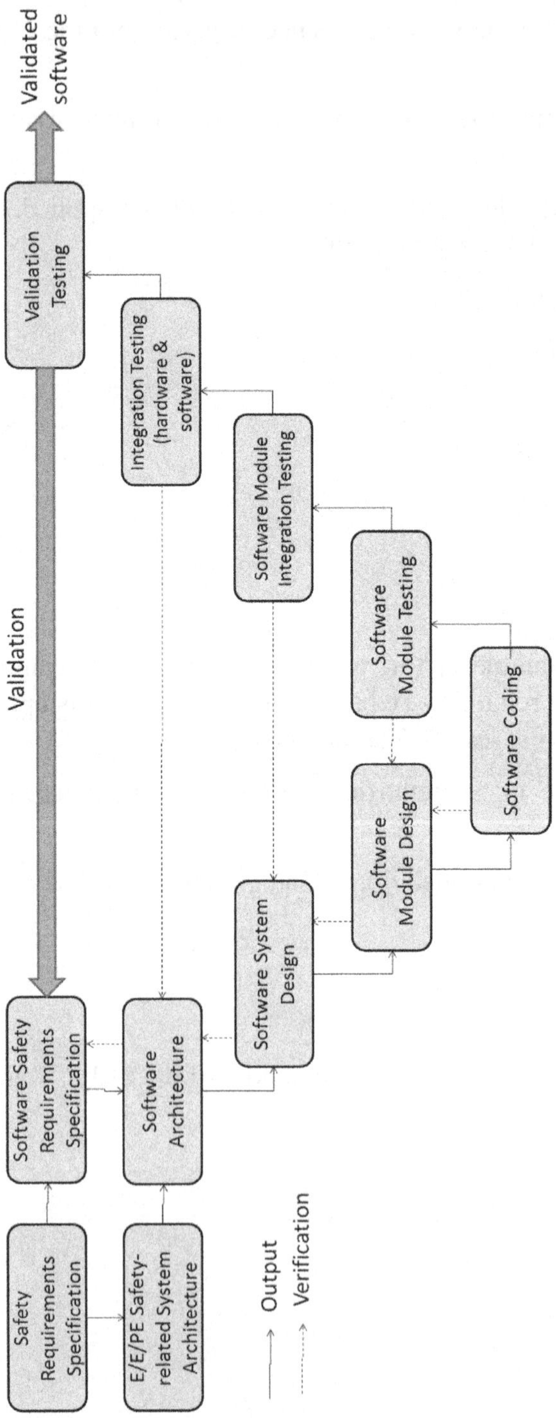

AS(IEC)61508.3

Table 7.8 provides details on the software development techniques and measures that are highly recommended by AS(IEC)61508.3 The purpose of this table is to draw the reader's attention to those elements of software design and development that are the most critical to compliance with AS(IEC)61508.3 for SIL1, SIL2 and SIL3. A range of other techniques and measures are recommended by AS(IEC)61508.3, but are arguably less critical than these. Designers should consult the standards and select techniques and measures, as appropriate to the application.

Table 7.8: AS(IEC)61508 'M' and 'HR' Techniques & Measures (Software)

Technique / Measure	AS(IEC)61508.7 or Other Reference	Interpretation of the Requirement	SIL1	SIL2	SIL3
Software Safety Requirements Specification					
Semi-formal methods.	AS(IEC)61508.3 Table B.7	Block diagrams, sequence diagrams, finite state machines, state transition diagrams, time petri nets, decision / truth tables.			Yes
Forward traceability between the system safety requirements and the software safety requirements.	C.2.11				Yes
Backward traceability between the system safety requirements and the software safety requirements.	C.2.11				Yes
Computer-aided specification tools to support appropriate techniques / measures above.	B.2.4	Tools supporting the chosen methods.			Yes
Software Design and Development - Software Architecture Design					
Fault detection.	C.3.1	Used as far as dealing with sensor, actuator and data transmission failures and which are not covered by the measures within the embedded system according to the requirements of AS(IEC)61508.2.			Yes
Graceful degradation.	C.3.8	Maintain more critical functions after failures, by dropping less critical functions.			Yes

Technique / Measure	AS(IEC)61508.7 or Other Reference	Interpretation of the Requirement	SIL1	SIL2	SIL3
Modular approach.	AS(IEC)61508.3 Table B.9	Software module size limit, information hiding / encapsulation, one entry / one exit point in sub-routines and functions, fully defined interfaces.	Yes	Yes	Yes
Use of trusted/verified modules and components.	C.2.10	SIL-certified function blocks and part programs.		Yes	Yes
Forward traceability between the system safety requirements and the software safety requirements.	C.2.11				Yes
Backward traceability between the system safety requirements and the software safety requirements.	C.2.11				Yes
Structured diagrammatic methods.	C.2.1	As per modular approach.	Yes	Yes	Yes
Semi-formal methods.	AS(IEC)61508.3 Table B.7	Block diagrams, sequence diagrams, finite state machines, state transition diagrams, time petri nets, decision / truth tables.			Yes
Computer-aided specification and design tools.	B.2.4	Tools supporting the chosen methods.			Yes
Cyclic behavior, with guaranteed maximum cycle time.	C.3.11			Yes	Yes
Time-triggered architecture.	C.3.11			Yes	Yes
Event-driven with guaranteed maximum response time.	C.3.11			Yes	Yes
Static resource allocation	C.2.6.3				Yes

Technique / Measure	AS(IEC)61508.7 or Other Reference	Interpretation of the Requirement	SIL1	SIL2	SIL3
Software Design and Development - Support Tools and Programming Language					
Suitable programming language.	C.4.5	Full variability, high level language. PLC supplier's proprietary language.	Yes	Yes	Yes
Strongly typed programming language.	C.4.1	IEC61131-1 structured text.	Yes	Yes	Yes
Language sub-set	C.4.2	Do not use complex 'macro' instructions or interrupts which can alter the PLC scan cycle etc…			Yes
Certified tools and certified translator.	C.4.3	As available from PLC supplier.		Yes	Yes
Tools and translator: Increased confidence from use.	C.4.4	PLC suppliers development kit and in-house tools developed over several projects.	Yes	Yes	Yes
Software Design and Development - Detailed Design					
Structured methods	C.2.1	Not used if limited variability programming applied.	Yes	Yes	Yes
Semi-formal methods.	AS(IEC)61508.3 Table B.7	Block diagrams, sequence diagrams, finite state machines, state transition diagrams, time petri nets, decision / truth tables.			Yes
Computer-aided design tools.	B.3.5	Used for the selected methods.			Yes
Defensive programming.	C.2.5	All measures except those which are automatically inserted by the compiler which are explicitly used in application software where they are effective.			Yes
Modular approach.	AS(IEC)61508.3 Table B.9	Software module size limit, information hiding / encapsulation, one entry / one exit point in sub-routines and functions, fully defined interfaces.	Yes	Yes	Yes
Design and coding standards	C.2.6 and AS(IEC)61508.3 Table B.1	Use of coding standard, no dynamic objects, no dynamic variables (or on-line checking), limited use of interrupts, limited use of pointers, limited use of recursion, no unconditional jumps in programs in higher level languages.		Yes	Yes
Structured programming.	C.2.7	As per modular approach.	Yes	Yes	Yes

Technique / Measure	AS(IEC)61508.7 or Other Reference	Interpretation of the Requirement	SIL1	SIL2	SIL3
Use of trusted/verified modules and components.	C.2.10	SIL-certified function blocks and part programs.		Yes	Yes
Software Design and Development - Software Module Testing and Integration					
Dynamic analysis and testing.	B.6.5 and AS(IEC)61508.3 Table B.2	Test case execution from boundary value analysis, structure-based testing.		Yes	Yes
Data recording and analysis	C.5.2	Records of test cases and results.	Yes	Yes	Yes
Functional and black box testing.	B5.1, B.5.2 and AS(IEC)61508.3 Table B.3	Boundary value analysis, equivalence classes and input partition testing.	Yes	Yes	Yes
Performance testing.	AS(IEC)61508.3 Table B.6	Avalanche / stress testing, response timings and memory constraints, performance requirements.			Yes
Interface testing.	C.5.3	Detect errors in interfaces of sub-programs.			Yes
Programmable Electronics Integration (Hardware and Software)					
Functional and black box testing.	B5.1, B.5.2 and AS(IEC)61508.3 Table B.3	Boundary value analysis, equivalence classes and input partition testing.	Yes	Yes	Yes
Performance testing.	AS(IEC)61508.3 Table B.6	Avalanche / stress testing, response timings and memory constraints, performance requirements.			Yes
Software Safety Validation					
Process simulation.	C.5.18				Yes
Modeling.	AS(IEC)61508.3 Table B.5	Finite state machines, performance modeling, time petri nets.			Yes
Functional and black box testing.	B5.1, B.5.2 and AS(IEC)61508.3 Table B.3	Boundary value analysis, equivalence classes and input partition testing.	Yes	Yes	Yes
Forward traceability between the system safety requirements specification and the software validation plan.	C.2.11				Yes

Technique / Measure	AS(IEC)61508.7 or Other Reference	Interpretation of the Requirement	SIL1	SIL2	SIL3
Backward traceability between the software validation plan and the software safety requirements specification.	C.2.11				Yes
Software Modification					
Impact Analysis	C.5.23	An impact analysis is carried out to consider how the effect of the proposed changes is limited by the modularity of the overall system.	Yes	Yes	Yes
Re-verify changed software modules.	C.5.23	Repeat earlier tests.	Yes	Yes	Yes
Re-verify affected software modules	C.5.23	Repeat earlier tests.		Yes	Yes
Re-verify complete system	AS(IEC)61508.3 Table A.7	Depending on the results of the impact analysis.			Yes
Regression validation	C.5.25			Yes	Yes
Software configuration management.	C.5.24	Baselines, records of changes, impacts on other system requirements.	Yes	Yes	Yes
Data recording and analysis	C.5.2	Records of test cases and results.	Yes	Yes	Yes
Forward traceability between the software safety requirements specification and the software modification plan.	C.2.11				Yes
Backward traceability between the software modification plan and the software safety requirements specification.	C.2.11				Yes

Technique / Measure	AS(IEC)61508.7 or Other Reference	Interpretation of the Requirement	SIL1	SIL2	SIL3
Software Verification					
Static analysis.	B.6.4 and AS(IEC)61508.3 Table B.8	Boundary value analysis, control flow analysis, data flow analysis, symbolic execution, walk-through / design reviews.		Yes	Yes
Dynamic analysis and testing.	B.6.5 and AS(IEC)61508.3 Table B.2	Test case execution from boundary value analysis, structure-based testing.		Yes	Yes
Forward traceability between the software design specification and the software verification plan.	C.2.11				Yes
Backward traceability between the software verification plan and the software design specification.	C.2.11				Yes
Offline numerical analysis	C.2.13				Yes
Functional Safety Assessment					
Failure analysis	AS(IEC)61508 Table B.4	Fault-tree analysis, failure modes, effects and criticality analysis.			Yes
Common cause failure analysis of diverse software	C.6.3	If diverse software is used.			Yes
Forward traceability between the requirements of AS61508-1 Clause 8 and the plan for software functional safety assessment.	C.2.11				Yes

AS(IEC)62061

AS(IEC)62061 Clause 6.10, 6.11 and 6.12 detail the requirements for avoidance and control of systematic failures for software used in E/E/PE safety-related systems, including the integration of software with hardware. Furthermore, Annex C provides guidelines for the design and development of embedded software[45]. The following is a summary of the requirements:

Software Safety Requirements Specification

1. Software safety requirements specification to be documented and made available to the designer.
2. Safety requirements for each sub-system to be specified, on the basis of:
    a. Specified requirements of each SRCF,
    b. Requirements resulting from the SRECS architecture,
    c. Any requirements of the Functional Safety Plan.
3. Specification to have sufficient detail to allow design to achieve the required SIL, and allow verification.
4. Software developer to review the specification to ensure the requirements are adequately specified.
5. Software requirements to be clear, verifiable, testable, maintainable and operable for the required SIL, traceable back to the SRECS safety requirements and free of ambiguous terminology and descriptions.
6. The requirements of each sub-system to be specified are:
    a. logic of all function blocks,
    b. I/O interfaces for each logic block,
    c. format and range values of I/O data,
    d. limits of each function block (eg. response time, limit values),
    e. diagnostic functions of other devices (eg. sensors) to be implemented by that sub-system,
    f. functions that enable the machine to achieve a safe state,
    g. functions related to detection, annunciation and handling of faults,
    h. functions related to periodic testing on-line and off-line,
    i. functions that prevent unauthorised modification,
    j. interfaces to non-SRCF's,
    k. capacity and response time performance.
7. Semi-formal methods (eg. logic, function-block or sequence diagrams) to be included in documentation.

Embedded Software Design and Development

1. Embedded software must comply with AS61508.3 as appropriate for the required SIL. There are two options for compliance:
    a. Use logic devices with previously certified embedded software, or
    b. Use AS62061 Annex C to establish the level of compliance.

---

[45] Software supplied by the manufacturer that is part of the SRECS and that is not normally accessible for modification. See Appendix A.

Application Software Design and Development (Full Variability Languages)

The same requirements should apply as for embedded software.

Application Software Design and Development (Limited Variability Languages)

1. General Requirements:
    a. The outcomes of the software development activities to be verified at appropriate stages,
    b. The design method and application language should facilitate modularity, functionality, information flow, sequencing, timing, data structures, comprehension by developers and others, verification and validation and safe modification. Where possible use previously certified design tools and application languages and previously certified or well-proven logic functions.
    c. Testing to be the main method of application software verification. A test plan to be developed,
    d. Where application software contains safety and non-safety related software, all software is to be treated as safety-related.
    e. Design to implement data integrity checks and reasonableness checks at the application layer.
    f. Design to implement self-monitoring of control flow and data flow unless already incorporated in the embedded software. On detection of software failure, action to be taken to achieve or maintain a safe state.
    g. The suitability of previously developed software library functions to be justified.
    h. Modification / changes to be subject to impact analysis that identifies all modules affected and necessary re-verification activities.
2. Software Configuration Management:
    a. Functional Safety Plan to define the strategy for development, integration, verification and validation of software.
    b. Software configuration management to ensure all necessary actions are carried out to achieve and maintain the required SIL, maintain identification of all documents related to change control, apply change control procedures, document the justification and approval of modifications, document the release of software versions and ensure maintenance of change control documentation for subsequent audit.
3. Software Architecture:
    a. To be based on the requirements of the overall safety requirements specification for the SRECS,
    b. Provide a complete description of the internal structure and operation of the SRECS and its components,
    c. Identify all software components and a description of their connection interactions with other components (hardware and software),
    d. Identify software modules included in SRECS but not used in safety-related operation.
    e. Techniques and measures for software design to be described and justified.
    f. Measures for maintaining integrity of data to be described and justified.

4. Support Tools, User Manual and Application Languages
    a. Suitable tools to be selected, explained and documented.
    b. Sub-sets of the application programming language to be described.
    c. Take account of constraints and known weaknesses included in the user manuals,
    d. Application language preferably to be completely and unambiguously defined,
    e. Procedures for application language to define good practice.
    f. Program documentation to include author, description, traceability to functional requirements, traceability to standard library functions, inputs and output and configuration management information.
5. Application Software Design:
    a. Ensure the following available prior to the start of design – software safety requirements specification, description of software architecture design, software validation plan,
    b. Structured design to be implemented, ie. modularity of functions and I/O, testability of functions, capacity for safe modification through traceability and explanation of application functions and constraints,
    c. Design of each module and the testing to be applied to be specified,
    d. Appropriate integration testing to be specified.
6. Application Code Development:
    a. Code to be readable, understandable, testable, and satisfy relevant design principles and safety requirements.
    b. Review code for conformance to specified design, coding rules and safety requirements.
7. Application Module Testing:
    a. Each I/O point to be checked to confirm it is mapped to the correct application logic.
    b. Each module to be checked to determine that intended function is correctly executed and unintended functions are not executed.
    c. Testing to include each branch, boundary values and sequences,
    d. Testing to be documented, including failures / non-conformances, corrective actions and re-testing,
    e. If a module has been previously assessed, the level of testing may be reduced.
8. Application Software Integration Testing
    a. Verify all modules interact correctly with each other and with the embedded software and perform their intended function and do not perform unintended functions,
    b. Testing to be documented, including failures / non-conformances, corrective actions and re-testing,
    c. Modification / changes during integration to be subject to impact analysis that identifies all modules affected and necessary re-verification activities.

Software-based Parameterisation

1. Parameterisation to be carried out using a tool provided by the supplier of the SRECS. The tool is to have its own identification (ie. name, version etc…). and is to have means of preventing unauthorised modification of parameters (eg. by password).
2. Integrity of parameterisation data to be maintained by controlling for invalid inputs, data corruption, incomplete transmission and failures in the parameterisation tool.
3. The parameterisation tool to meet all relevant requirements of AS62061 for a sub-system, have a special procedure for setting of safety-related parameters and diverse software modules to be employed for encoding/decoding and data transmission functions.
4. Document the parameters associated with the SRECS, the parameter values entered and the person responsible for the parameterisation.
5. Verify correct settings, plausibility of values, no unauthorised modifications and correct data transmission.

SRECS Integration and Testing - General Requirements

1. Integrate according to the specified SRECS design and integration tests.
2. Verify all modules perform their intended function and do not perform unintended functions,
3. Testing to ensure compatibility of the application software with the hardware and the embedded software platform,
4. Testing to be documented, including failures / non-conformances, corrective actions and re-testing,
5. Modification / changes during integration to be subject to impact analysis that identifies all components affected and necessary re-verification activities.
6. Test records to include version of test specification used, acceptance criteria, version of SRECS tested, tools and equipment used, calibration data, result of each test, discrepancies observed and decisions taken on whether to continue test or issue change request.

SRECS Integration and Testing - Tests to Determine Systematic Safety Integrity

1. Functional and proof testing – initiate the function and record the response of the function and its sub-systems.
2. Dynamic testing – test the behaviour of the SRECS under fault and in realistic functional conditions (eg. fault insertion testing).
3. Review test results to see whether specified characteristics of the safety requirements specification of the SRECS have been achieved.

ISO13849

ISO13849-1 Section 4.6 and Annex J details the requirements for avoidance and control of systematic failures for software used in SRP/CS. The following is a summary of the requirements:

Embedded Software Design and Development

For PLa to PLd, the following basic measures should be applied:

1. Software safety lifecycle with verification and validation activities, per Figure 7.22.
2. Documentation of specification and design.
3. Modular and structured design and coding.
4. Control of systematic failures per Annex G.2 (see Section 7.15).
5. Where using software-based measures for control of random hardware failures, verification of correct implementation.
6. Functional testing, e.g. black box testing.
7. Appropriate software safety lifecycle activities after modifications.

For PLc and PLd, the following additional measures should be applied:

1. Project management and quality management system comparable to, e.g. AS/IEC 61508 or ISO9001.
2. Documentation of all relevant activities during software safety lifecycle.
3. Configuration management to identify all configuration items and documents related to an embedded software release.
4. Structured specification with safety requirements and design.
5. Use of suitable programming languages and computer-based tools with confidence from use.
6. Modular and structured programming, separation in non-safety-related software, limited module sizes with fully defined interfaces, use of design and coding standards.
7. Coding verification by walk-through/review with control flow analysis.
8. Extended functional testing, e.g. grey box testing, performance testing or simulation.
9. Impact analysis and appropriate software safety lifecycle activities after modifications.

For PLe, AS(IEC) 61508.3 should also apply, as appropriate for SIL3. Alternatively, if redundant channels are to be used and 'diversity' principles are applied in their specification, design and coding, PLe can be achieved overall if the above-mentioned measures for PLc and PLd are applied for each of the redundant channels.

Application Software Design and Development (Full Variability Languages)

The same requirements should apply as for embedded software.

Application Software Design and Development (Limited Variability Languages)

Note: If a part of the application software within one component has any impact on several safety functions with different PL, then the requirements related to the highest PL shall apply.

For PLa to PLe, the following basic measures should be applied:

1. Development lifecycle with verification and validation activities, see ISO13849-1 Figure 1.
2. Documentation of specification and design.
3. Modular and structured programming.
4. Functional testing.
5. Appropriate development activities after modifications.

For PLc to PLe, the following additional measures should be applied (ie. recommended for PLc, highly recommended for PLd, mandatory for PLe):

1. The safety-related software specification should be reviewed, made available to every person involved in the lifecycle and should contain the description of:
    a. Safety functions with required PL and associated operating modes.
    b. Performance criteria, e.g. reaction times.
    c. Hardware architecture with external signal interfaces.
    d. Detection and control of external failure.
2. Selection of tools, libraries, languages.
    a. Suitable tools with confidence from use.
    b. Validated function block libraries should be used.
    c. Graphical languages (e.g. function block diagram, ladder diagram) are highly recommended.
3. Software design should feature:
    a. Semi-formal methods to describe data and control flow, e.g. state diagram or program flow chart.
    b. Modular and structured programming predominantly realized by function blocks deriving from safety-related validated function block libraries.
    c. Function blocks of limited size of coding.
    d. Code execution inside function block which should have one entry and one exit point.
    e. Architecture model of three stages, ie. Inputs, Processing, Outputs.
    f. Assignment of a safety output at only one program location.
    g. Use of techniques for detection of external failure and for defensive programming within input, processing and output blocks which lead to safe state.
4. Where safety and non-safety application software are combined in one component:
    a. They should be coded in different function blocks with well-defined data links;

b. There should be no logical combination of non-safety-related and safety-related data which could lead to downgrading of the integrity of safety-related signals, for example, combining safety-related and non-safety-related signals by a logical "OR" where the result controls safety-related signals.

5. Software implementation/coding:
    a. Code should be readable, understandable and testable and, because of this symbolic variables (instead of explicit hardware addresses) should be used.
    b. Justified or accepted coding guidelines should be used.
    c. Data integrity and plausibility checks (e.g. range checks.) available on application layer (defensive programming) should be used.
    d. Code should be tested by simulation.
    e. Verification should be by control and data flow analysis for PLd and PLe.
6. Testing:
    a. The appropriate validation method is black-box testing of functional behaviour and performance criteria (e.g. timing performance).
    b. Test planning is recommended and should include test cases with completion criteria and required tools.
    c. I/O testing should ensure that safety-related signals are correctly used within the application software.
    d. For PLd and PLe, test case execution from boundary value analysis is recommended.
7. Documentation
    a. All lifecycle and modification activities shall be documented.
    b. Documentation shall be complete, available, readable and understandable.
    c. Code documentation within source text shall contain module headers with legal entity, functional and I/O description, version and version of used library function blocks, and sufficient comments of networks/statement and declaration lines.
8. Verification – by review, inspection, walkthrough or other appropriate activities. Verification is only necessary for application code, not for validated library functions.
9. Configuration management - it is highly recommended that procedures and data backup be established to identify and archive documents, software modules, verification/validation results and tool configuration related to a specific application software version.
10. Modifications - after modifications of application software, impact analysis shall be performed to ensure specification. Appropriate lifecycle activities shall be performed after modifications. Access rights to modifications shall be controlled and modification history shall be documented.

Software-based Parameterisation

The same requirements apply as AS(IEC)62061.

7.17 AS(IEC)61508 Phase 10 – Route 2 and 'Proven-In-Use'

The 2010/2011 issue of AS(IEC)61508.2 provides a revised concept for complying with the requirements for hardware integrity and systematic failure avoidance and control.

In AS61508.2 Clause 7.4.2.2 three (3) options are provided for compliance with systematic failure avoidance and control requirements:

Route 1_S: compliance with the established deterministic requirements for the avoidance and control of systematic faults.

Route 2_S: compliance with the requirements for evidence that the equipment is "proven-in-use".

Route 3_S: (for pre-existing software elements only): compliance with the requirements of AS(IEC)61508.3 Clause 7.4.2.12.

In AS61508.2 Clause 7.4.4 two (2) options are provided for compliance with hardware safety integrity requirements:

Route 1_H: based on the established hardware fault tolerance and safe failure fraction concepts.

Route 2_H: based on component reliability data from field feedback from end-users, increased confidence levels and specified hardware fault tolerance for specified safety integrity levels.

The previous sections are based on the use of Routes 1_S and 1_H (the established approach). This section provides guidance on the use of Route 2_S and 2_H.

Route 2_S Requirements

The requirements for evidence that an item is "proven-in-use" are provided in AS(IEC)61508.2 Clause 7.4.10. There are seven (7) requirements to be met, as follows:

7.4.10.1 *An element shall only be regarded as proven in use when it has a clearly restricted and specified functionality and when there is adequate documentary evidence to demonstrate that the likelihood of any dangerous systematic faults is low enough that the required safety integrity levels of the safety functions that use the element is achieved. Evidence shall be based on analysis of operational experience of a specific configuration of the element together with suitability analysis and testing.*

7.4.10.2 *The documentary evidence required by 7.4.10.1 shall demonstrate that:*

a) the previous conditions of use of the specific element are the same as, or sufficiently close to, those that will be experienced by the element in the E/E/PE safety-related system;

b) the dangerous failure rate has not been exceeded in previous use.

7.4.10.3 *When there is any difference between the previous conditions of use and those that will be experienced in the E/E/PE safety-related system, then an impact analysis on the differences shall be carried out using a combination of appropriate analytical methods and testing, in order to demonstrate that the likelihood of any dangerous systematic faults is low enough that the required safety integrity level(s) of the safety function(s) that use the element is achieved.*

***7.4.10.4** A proven in use safety justification shall be documented, using the information available from 7.4.10.2, that the element supports the required safety function with the required systematic safety integrity. This shall include:*

a) the suitability analysis and testing of the element for the intended application;

b) the demonstration of equivalence between the intended operation and the previous operation experience, including the impact analysis on the differences;

c) the statistical evidence.

***7.4.10.5** The following factors shall be taken into account when determining whether or not the above requirements (7.4.10.1 to 7.4.10.4) have been met, in terms of both the coverage and degree of detail of the available information:*

a) the complexity of the element;

b) the systematic capability required for the element;

c) the novelty of design.

***7.4.10.6** There shall be satisfactory evidence that, the existing element's functions that are not covered by the proven in use demonstration, cannot adversely affect the safety integrity of the element functions that are used.*

***7.4.10.7** Any future modification of a proven in use element shall comply with the requirements of 7.8, and IEC 61508-3.*

Clause 7.4.10.1 defines the conditions in which the Route 2_S "proven-in-use" approach may be taken. Clauses 7.4.10-2 to 7.4.10-7 provide conditions and rules regarding the "proven-in-use" justification.

Clause 7.4.10.1 specifies "proven-in-use" is only available if the element:

- Has clearly specified and restricted functionality, and
- Has adequate documentary evidence that the likelihood of dangerous systematic faults is low enough for the required SIL of the safety functions that use it, determined from operational experience of the element in a specific configuration suitability analysis and testing.

The first point means that Route 2_S would usually be restricted to the devices or sub-systems, rather than overall safety-related systems. Devices and sub-systems are more likely to have clearly specified and restricted functionality, for example, a sensor, protection relay, contactor or valve.

The second point means that some form of analysis must be performed to quantify the probability of systematic failure of the device, so that it can be compared with the probability of failure of the required SIL.

It is therefore typical in the "proven-in-use" approach that a statistical analysis of field failures occurs. From this analysis, failure probabilities are derived and compared to the required SIL's. This statistical analysis of field failures aligns with and complements the Route 2_H approach for complying with the hardware integrity requirements of AS(IEC)61508. It is often the case that when Route 2_S is selected, Route 2_H is also selected.

Route 2_H Requirements

The requirements are provided in AS(IEC)61508.2 Clause 7.4.4.3. There are four (4) requirements to be met, as follows:

7.4.4.3.1 *The minimum hardware fault tolerance for each subsystem of an E/E/PE safety-related system implementing a safety function of a specified safety integrity level shall be as follows:*

a) a hardware fault tolerance of 2 for a specified safety function of SIL 4 unless the conditions in 7.4.4.3.2 apply.

b) a hardware fault tolerance of 1 for a specified safety function of SIL 3 unless the conditions in 7.4.4.3.2 apply.

c) a hardware fault tolerance of 1 for a specified safety function of SIL 2, operating in a high demand or continuous mode of operation, unless the conditions in 7.4.4.3.2 apply.

d) a hardware fault tolerance of 0 for a specified safety function of SIL 2 operating in a low demand mode of operation.

e) a hardware fault tolerance of 0 for a specified safety function of SIL 1.

7.4.4.3.2 *For type A elements only, if it is determined that by following the HFT requirements specified in 7.4.4.3.1, for the situation where an HFT greater than 0 is required, it would introduce additional failures and lead to a decrease in the overall safety of the EUC, then a safer alternative architecture with reduced HFT may be implemented. In such a case this shall be justified and documented. The justification shall provide evidence that:*

a) compliance with the HFT requirements specified in 7.4.4.3.1 would introduce additional failures and lead to a decrease in the overall safety of the EUC; and

b) if the HFT is reduced to zero, the failure modes, identified in the element performing the safety function, can be excluded because the dangerous failure rate(s) of the identified failure mode(s) are very low compared to the target failure measure for the safety function under consideration (see 7.4.4.1.1 c)). That is, the sum of the dangerous failure frequencies of all serial elements, on which fault exclusion is being claimed, should not exceed 1 % of the target failure measure. Furthermore the applicability of fault exclusions shall be justified considering the potential for systematic faults

7.4.4.3.3 *If Route 2H is selected, then the reliability data used when quantifying the effect of random hardware failures (see 7.4.5) shall be:*

a) based on field feedback for elements in use in a similar application and environment; and,

b) based on data collected in accordance with international standards (e.g., IEC 60300-3-2 or ISO 14224:); and,

c) evaluated according to:

> *i) the amount of field feedback; and,*
>
> *ii) the exercise of expert judgement; and where needed, iii) the undertaking of specific tests; in order to estimate the average and the uncertainty level (e.g., the 90 % confidence interval or the probability distribution (see Note 2)) of each reliability parameter (e.g., failure rate) used in the calculations.*

7.4.4.3.4 *All type B elements used in Route 2H shall have, as a minimum, a diagnostic coverage of not less than 60 %.*

Route 2_H requires that failure rate estimates are provided at the 90% confidence interval. AS(IEC)61508.2 Clause 7.4.9.5 Note 2 indicates that these may be determined using the χ^2 (chi-square) function or via a Bayesian approach.

A common approach is to perform a "single-sided" χ^2 estimate. The statistical interpretation of a 'single-sided" χ^2 failure rate estimate at 90% confidence is as follows:

"We are 90% confident that the true failure rate is not higher than ..."

Further information on the χ^2 failure rate estimation method may be obtained online at: http://www.itl.nist.gov/div898/handbook/apr/section4/apr451.htm

A simplified equation for a "single-sided" χ^2 failure rate estimate (on a "per item" basis) at 90% confidence is:

$$\lambda_{90\%} = \frac{\chi^2_{n=2k+2,\alpha=0.1}}{2T}$$

where:

χ^2 = represents a multiplier determined from the "chi-square" function[46]

k = the number of failures experienced

α = 1- confidence level (ie. 1 - 90% = 0.1)

T = accumulated time in use for all devices (ie. number of devices x time in use)

For example, consider a population of 100 identical items in service in similar applications and environments. Suppose the items are fielded for 1 year and only 1 failure is experienced within the population of items during that time.

The "single-sided" χ^2 failure rate estimate, on a "per item" basis, at 90% confidence is:

$$\lambda_{90\%} = \frac{\chi^2_{n=2k+2,\alpha=0.1}}{2T} = \frac{\chi^2_{n=4,\alpha=0.1}}{2x100x1} = \frac{7.78}{200} = 0.039\, failures/yr = 4.4x10^{-6}\, failures/yr$$

It is important to recognize that a χ^2 failure rate estimate is not an "average". The "average" failure rate on a "per item" basis would be calculated as:

λ_{ave} = 1 failure / (100 items x 1 year) = 0.01 failures per year.

[46] This is usually determined by referring to mathematical tables or the use of the 'chi-square' function provided on some scientific calculators or in spreadsheet software such as MSExcel.

A "single-sided" χ^2 estimate may also be derived in cases where no failures have been recorded.

A simplified equation for a no-failures "single-sided" χ^2 failure rate estimate (on a "per item" basis) at 90% confidence is:

$$\lambda_{90\%} = \frac{2.3}{T}$$

where:

T = accumulated time in use for all devices (ie. number of devices x time in use)

For example, consider a population of 100 identical items in service in similar applications and environments. Suppose the items are fielded for 1 year and no failures are experienced in that time.

The "single-sided" χ^2 failure rate estimate, on a "per item" basis, at 90% confidence is:

$$\lambda_{90\%} = \frac{2.3}{T} = \frac{2.3}{100x1} = 0.023\, failures/\,yr = 2.6x10^{-6}\, failures/\,hr$$

The "average" failure rate on a "per item" basis would be calculated as:

λ_{ave} = 0 failures / (100 items x 1 year) = 0 failures per year.

7.18 AS(IEC)61508 Phase 11 – "Other Technology" Safety-related Systems Specification and Realisation

Note: See Section 4.4 regarding the validity of the design and development approach provided in Part 1501 and 1502 of this standard. It is my view that the approach to the specification and realisation of machinery-sector safety-related systems provided by the latest issues of ISO13849 or AS(IEC)62061 should be used, and the out-dated approaches provided AS4024.1501-2006 and AS4024.1502-2006 should be avoided where possible. AS4024.1 was in the process of being updated by Standards Australia at the time of writing of this book. It is anticipated that AS4024.1 may be re-issued in 2013 or 2014 taking account of the latest issues of ISO12100, ISO14121 and ISO13849.

Objective/s:

"Other technology" safety-related systems are included within the definition of "other risk reduction measures" but are not within the scope of AS(IEC)61508 or AS(IEC)62061.

Therefore, an alternate design and development process needs to be undertaken in lieu of AS(IEC)61508 / AS(IEC)62061.

Design and development of "other technology" (ie. mechanical, pneumatic, hydraulic etc...) safety-related systems should follow the process provided by ISO-13849-1 Clause 4.4 and Figure 3. These systems are not within the scope of AS(IEC)61508, AS(IEC)62061 or AS(IEC)61511, however in AS(IEC)62061 reference is made to ISO13849-1 and ISO13849-2 for design and development of these systems.

Responsible Entity:

Whilst ultimate responsibility for ensuring completion of the requirements of these phases rests with the end-user (eg. the mining company), responsibility for conducting the activities may be delegated to the vendor or designer.

Activities to be Undertaken:

ISO13849-1 and ISO13849-2 should be applied. Each "other technology" safety function should be specified and designed to meet the requirements of an appropriate Performance Level (ie. PL a, b, c, d, e). Where "other technology" parts are used in conjunction with an E/E/PE safety-related system these should be designed to meet the PL that is equivalent to the SIL that was allocated to the E/E/PE safety-related system. . ISO13849-1 Table 4 provides an equivalence of PL and SIL levels for this purpose (as shown in Table 5.2).

Documentation to be Kept:

The documentation required depends upon the PL specified for each safety function, but the following should be used as a guideline:

1. Sufficient documentation should be generated that describes the design of the "other technology" safety-related system and the achievement of its specific requirements. The list of 'information for use' in ISO13849-1 Clauses 10 and 11 provide a reasonable guideline for this.
2. Design and verification documentation for each safety function, per ISO13849-1 Clause 4.
3. Documentation for validation phase for each safety function, per ISO13849-2.

7.19 AS(IEC)61508 Phase 11 – "Other Risk Reduction Measures" Specification and Realisation

Objective/s:

Design guidelines for "other risk reduction measures" are not within the scope of AS(IEC)61508. Design guidelines need to be applied which are appropriate to type and the application.

Responsible Entity:

Whilst ultimate responsibility for ensuring completion of the requirements of these phases rests with the end-user (eg. the mining company), responsibility for conducting the activities may be delegated to the vendor or designer.

Activities to be Undertaken:

The appropriate standards and guidelines pertaining to the design of each type of "other risk reduction measure" should be identified and applied early in the realisation stage.

For example, fixed and movable guards which are used on machinery may be designed to AS4024.1 Part 1601 and any interlocking devices to be used with guards may be designed to AS4024.1 Part 1602.

Documentation to be Kept:

Sufficient documentation should be generated that describes the design of each "other risk reduction measure" and the achievement of its specific requirements, eg. for fixed or movable guards - sufficient evidence to demonstrate compliance to AS4024.1 Part 1601.

7.20 AS(IEC)61508 Phase 12 - Overall Installation and Commissioning

Objective/s:

AS(IEC)61508 Phase 12 is primarily concerned with installing and commissioning the E/E/PE safety-related systems. AS(IEC)62061 Clause 13 also provides guidance for machinery-based applications.

Responsible Entity:

Whilst ultimate responsibility for ensuring completion of the requirements of these phases rests with the end-user (eg. the mining company), responsibility for conducting the activities may be delegated to the vendor or designer.

Activities to be Undertaken:

From AS(IEC) 61508-1 Clause 7.13.2, installation and commissioning of the E/E/PE safety-related systems should comply with the following, however the scope should be broadened to include the "other technology" safety-related systems and the "other risk reduction measures".

1. Installation and commissioning must be carried out in accordance with the Installation and Commissioning Plan developed during Phase 8

Documentation to be Kept:

The information and results acquired should be maintained throughout the overall safety lifecycle. The Safety Lifecycle Dossier should be updated with the following information:

1. Installation and Commissioning Report, including
a. The installation activities and resolution of failures and incompatibilities
b. The commissioning activities and resolution of failures and incompatibilities, along with references to failure reports

7.21 AS(IEC)61508 Phase 13 - Overall Safety Validation

Objective/s:

The objective of AS(IEC)61508 Phase 13 is to validate that the E/E/PE and "other technology" safety-related systems and the "other risk reduction measures" (eg. guards, barriers, procedures etc...). This is to be conducted on as 'as-implemented' basis against the requirements that were specified in the risk assessment and Safety Requirements Specification (SRS) developed during Phases 1-5. AS(IEC)62061 Clause 8 also provides guidance for machinery-based applications.

'Validation' is different in nature to 'verification'. AS(IEC)61508-4 describes the difference, as follows:

Validation – confirmation by examination and provision of objective evidence that the particular requirements for the specific intended use are fulfilled.

Verification – confirmation by examination and provision of objective evidence that the requirements have been fulfilled.

Responsible Entity:

Whilst ultimate responsibility for ensuring completion of the requirements of these phases rests with the end-user (eg. the mining company), responsibility for conducting the activities may be delegated to the vendor or designer.

Activities to be Undertaken:

From AS(IEC)61508.1 Clause 7.14 (using the information sourced in AS(IEC) Phases 4 and 5 and AS(IEC)62061 Steps 1-4), the following should occur:

1. Validation should be carried out in accordance with the Validation Plan prepared in accordance with the Functional Safety Management Plan,
2. All equipment used for quantitative measurements as part of the validation activities should be calibrated against a specification traceable to a national standard or to a vendor specification,
3. When discrepancies occur between actual and expected results, the analysis made and the decisions taken on whether to continue the validation or issue a change request and return to an earlier part of the validation should be documented.

This step itself comprises an overall validation that the safety requirements have been met prior to hand-over of the EUC, ensures that all safety-related systems and other risk controls are installed and operating correctly and confirms that all the necessary procedures and arrangements are in place to ensure that the functional safety of the EUC can be maintained through life.

Documentation to be Kept:

The information and results acquired should be maintained throughout the overall safety lifecycle. The Safety Lifecycle Dossier should be updated with the following information:

1.	Identification of the version of the SRS under which the validation was carried out
2.	Identification of the safety functions being validated (by test or analysis)
3.	Tools and equipment used with calibration data
4.	Documentation of the validation activities
5.	The results of the validation activities, including recorded discrepancies between expected and actual results
6.	Configuration identification of the item under test, the procedures applied and the test environment
7.	A statement of the validator's competence and independence

7.22 Beyond Pike River – Designing for "Survivable" Systems

> *Installation of explosion protection by means of a bypass in the underground workings near the fan 'may have contributed to reducing the extent or even the level of damage to the mine; as well as providing potential* ***survivability*** *of the ventilation system for later operation' (page 94).*
>
> *This incident provided a warning of a major hazard. It demonstrated the* ***vulnerability*** *of the mine's stoppings, as well as the potential for a roof fall in the hydro goaf to damage the ventilation system and lead to an explosive accumulation of methane (page 102).*
>
> *Second, the incident highlighted the* ***vulnerability*** *at the mine caused by Pike's lack of a full-time ventilation engineer (page 103).*
>
> *The collapse of the stopping at cross-cut 1 in the hydro panel on 30 October 2010 had served as a warning of this* ***vulnerability*** *(page 103).*
>
> *They remained working in the explosive atmosphere for at least 10 minutes, and 'there was no investigation because I never reported it'. He said there were times when they continued working in 2% methane contrary to the regulations. On another occasion, three contractors were found working in the ventilation return without gas detectors. These examples demonstrate the* ***vulnerability*** *of any system that simply assumes workers will comply with procedures, even those of such importance (page 143).*
>
> ***Royal Commission on the Pike River Coal Mine Tragedy, Volume 2, 2012.***

The words "vulnerability" and "survivability" appear often in reports about major accidents. The ability of equipment installed in a facility to survive a natural (eg. earthquake, cyclone etc...) or man-made (eg. gas explosion, fire, mud-rush etc...) disturbance can have a significant impact on the recovery process and the ability of exposed workers to survive.

The Pike River coal mine explosion in New Zealand on 19th November 2010 in which 29 workers died serves as a recent example. At Pike River the main ventilation fan was located underground. This was the only coal mine in the world where a main ventilation fan had been located underground. The fan was rendered inoperable during the explosion and main ventilation was unable to be restored to the mine.

The Royal Commission report noted the following:[47]

Three main risks arise from locating a main fan underground. First, it is more difficult to re-establish ventilation after an explosion, which could compromise the survival chances for anyone underground. Second, an underground fan is more likely to be damaged by an explosion. Third, an underground fan is closer to explosive material such as methane and coal dust, and a malfunction of the fan or its motor can be a source of ignition.

In terms of AS(IEC)61508 ventilation is an "other risk reduction measure" against the hazard of an explosive air / methane mixture. However, there are also safety-related systems used against the same hazard. For instance, a mine's gas monitoring systems are essential during the normal operation of a mine, but will also be critical to establishing the state of the mine's atmosphere and for rescue planning after a fire or explosion.

Not only do such systems need to operate correctly during normal operation, they should be able to be rapidly restored to service after a disturbance to assist the recovery process. They need to be designed for "survivability'.

---

[47] *Royal Commission on the Pike River Coal Mine Tragedy*, Volume 2, 2012, page 84.

"Survivability" is the ability of an entity - a system, sub-system, equipment, process, or a procedure - to continue to function during and after a natural or man-made disturbance. Survivability is usually qualified by specifying the range of conditions over which the entity will survive, the minimum acceptable level of post-disturbance functionality and the maximum acceptable outage duration.

"Survivability" is therefore an issue that potentially affects all equipment in a facility, whether it is an EUC, an EUC control system, a safety-related system or an "other risk reduction measure".

The concept of "survivability" has evolved from military environments, where it has been defined as the ability to remain "mission capable" after a single engagement. In this context survivability was seen to be affected by the following three (3) factors:

- **Susceptibility** - the inability to avoid being hit (by a weapon)
- **Vulnerability** - the inability to withstand the hit
- **Recoverability** - longer term post hit effects, damage control and firefighting, capability restoration, escape and evacuation

In a more general sense, "survivability engineering" focuses on preserving essential functions when systems are compromised. For a system to survive, it must recover from a damaging effect long before the underlying cause is identified. In fact, recovery must be successful whether or not the cause is ever determined.

Survivability engineering is therefore focused on (3) key design strategies, the "3R's":

- **Resistance** – the inherent capability of a system to repel disturbances
- **Recognition** – the capability to detect disturbances as they occur and to evaluate the extent of damage and compromise.
- **Recovery** – the capability to maintain essential services and assets during the disturbance, limit the extent of damage, and restore full services following the disturbance.

AS(IEC)61508 does not address "survivability" beyond the potential vulnerability of communication networks to security threats. AS(IEC)61508.1 states in Clause 7.5.2.2:

7.5.2.2 If security threats have been identified, then a vulnerability analysis should be undertaken in order to specify security requirements.

In the light of the Pike River and other disasters it is clear that a more comprehensive approach is needed - addressing the survivability of the EUC, EUC control system, safety-related systems and "other risk reduction measures", regardless of the type of technology, against a range of credible disturbances.

Survivability engineering approaches have been proposed to address the survivability of systems in information technology, military and aerospace applications. Examples are:

- DI ENVR 80266:86 : Nuclear Hardness And Survivability Design Analysis Report
- MIL HDBK 1799:1997 : Survivability, Aeronautical Systems (For Combat Mission Effectiveness)
- MIL HDBK 2069:1997 : Aircraft Survivability
- IEC 62443: Industrial communication networks - Network and system security

With the exception of IEC62443, these standards would generally not be applicable to mining or industrial environments.

Survivability engineering and analysis is an emerging field and a number of methodologies have been proposed. A detailed survey of this subject is beyond the scope of this book, but a common theme of several of these methodologies is that they are "risk based", that is:

1. The credible natural and man-made disturbances that may damage / compromise the operation of systems are identified.
2. The range of conditions over which the system must survive are defined.
3. The minimum acceptable level of post-disturbance functionality and the maximum acceptable outage duration are defined.
4. The systems that may be damaged / compromised are identified for each credible disturbance for which survival is required.
5. The reasons for system damage / compromise are identified (ie. susceptibilities and vulnerabilities).
6. The effects of the damage / compromise to those systems and their ability limit and / or overcome damage / compromise are assessed. (ie. recoverability).
7. Strategies are identified to minimise susceptibilities and vulnerabilities and maximize recoverability, using the principles embodied in the "3R's" – Resistance, Recognition, Recovery.

Survivability strategies should be developed for systems with reference to the requirements of the Model Act and Regulations, such as the use of the hierarchy of risk controls and the consideration of all known, available and suitable risk controls to satisfy the PCBU's primary obligation[48].

[48] See Chapter 3.

7.23 AS(IEC)61508 Phase 14 - Operation and Maintenance

Objective/s:

AS(IEC)61508 Phase 14 is primarily concerned with operating and maintaining the E/E/PE, "other technology" safety-related systems and "other risk reduction measures" in order that the functional safety of the EUC is maintained. AS(IEC)62061 also provides guidance for this phase, so a blend of the requirements from both standards is given here.

Responsible Entity:

Whilst ultimate responsibility for ensuring completion of the requirements of these phases rests with the end-user (eg. the mining company), responsibility for conducting the activities may be delegated to vendors, contractors and others.

Activities to be Undertaken:

From AS(IEC)61508.1 Clause 7.15 the following should occur:

1. Preparation of plans and procedures for maintaining the safety-related systems; including operation, maintenance and repair procedures for hardware and software
2. Implementation of plans and procedures
3. Checking compliance to maintenance schedules
4. Maintaining documentation
5. Periodic functional safety audits
6. Documenting any modifications that are made

Compliance to plans, procedures, maintenance schedules and the effective management of the configuration of the EUC, the EUC control system and the safety-related systems will be paramount to the continued achievement of the functional and safety integrity requirements. All of the procedures, training and arrangements necessary to ensure that the functional and safety integrity requirements of the safety-related systems will continue to be met must be documented, communicated, implemented, monitored, regularly reviewed and continuously improved. Figure 7.23 shows a suggested flow-chart for activities to be undertaken during this phase. This flowchart has been derived from AS(IEC)61508.1 Figures 7 and 8.

Figure 7.23: Operations & Maintenance Phase Activities

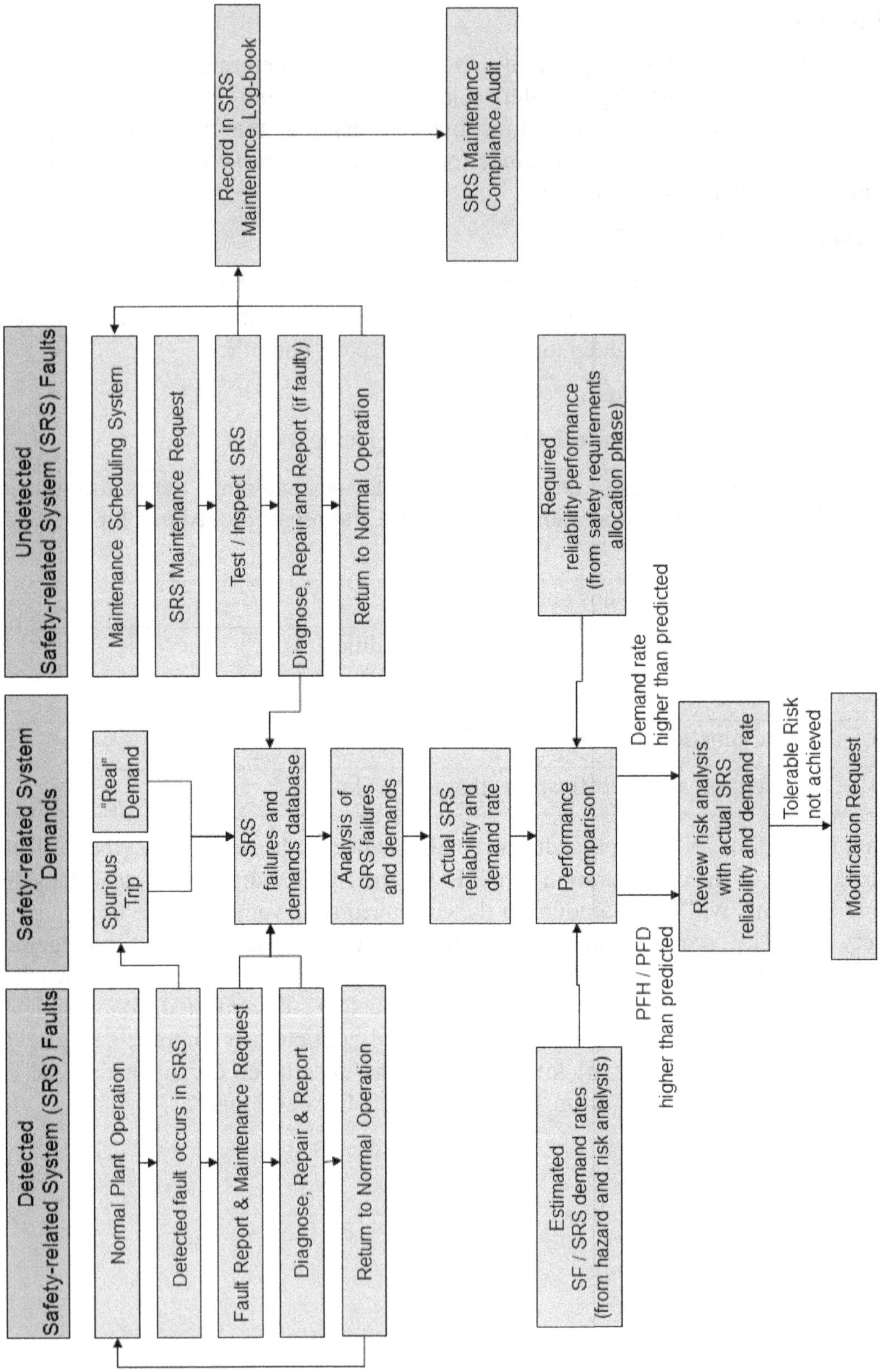

This flow-chart is primarily focused on the resolution of detected and undetected faults in the Safety Functions through-life, in order to assure reliability. Importantly, the actual reliability performance of the Safety Functions should eventually be determined and compared with earlier predictions so that action make be taken to alleviate any discrepancy.

The flow-chart also takes account of maintenance compliance to the specified Safety Function proof-testing intervals, as this may be a key determinant of Safety Function reliability.

If the specified PFH / PFD targets are not achieved in operation, or the actual SRCF demand rate increases beyond that expected, there are potential ramifications. These factors may impact on the ability of the installed risk controls to achieve the targeted tolerable level of risk. In this event, there may be a need for system modifications, which may include the EUC, the EUC control system, the safety-related systems and the "other risk reduction measures" – leading to the initiation of AS(IEC)61508 Phase 15.

Audits of functional safety performance should be carried out regularly, in a similar fashion to quality audits that are carried out on other management systems. For the E/E/PE and "other technology" safety-related systems, these audits should focus on the following issues:

1.	Continued implementation of operation and maintenance procedures
2.	Compliance to maintenance schedules
3.	Competence levels of those involved in operating and maintaining the Safety-related Systems
4.	Maintenance of the Safety Lifecycle Dossier
5.	Evidence of recording and resolution of functional and safety integrity (reliability) issues encountered with the safety-related systems in operation
6.	Whether performance comparisons have been carried out between the SIL / CAT actually achieved by each Safety Function and that predicted during the design phase
7.	Compliance to the Configuration (Engineering Change) Management Plan (see AS(IEC)61508 Phase15) when modifications and retrofits are carried out on the EUC, EUC control system, the safety-related systems or the "other risk reduction measures"

Internal auditing against the requirements for operation and maintenance of the safety-related Systems against their functional safety requirements should be carried out at regular intervals and corrective action taken, where appropriate.

Information for installation, use and maintenance should be provided by the designer in accordance with AS(IEC)62061 Clause 7. The scope of this information is provided in Section 6.4.

Documentation to be Kept:

The information and results acquired should be maintained throughout the overall safety lifecycle. The Safety Lifecycle Dossier should be updated with the following information:

1. Operation & Maintenance manuals
2. Register of engineering changes made to hardware and software
3. Register of changes to technical documentation
4. The results of any functional safety audits
5. Safety-related systems failures and incidents
6. Root Cause Analysis investigations into failures and incidents
7. Evidence of regular checking of maintenance compliance
8. Evidence of compliance to the Functional Safety Management Plan, Configuration Management Plan and other associated plans and procedures via Quality Management audits

7.24 AS(IEC)61508 Phase 15 - Modification and Retrofit

Objective/s:

AS(IEC)61508 Phase 15 is primarily concerned with ensuring that modifications are conducted in such a way that functional safety is maintained. AS(IEC)62061 and AS(IEC)61508.1 Clause 7.16 require that a Configuration (Engineering Change) Management Plan be produced. AS(IEC)62061 also provides guidance for this phase, so a blend of the requirements from both standards is given here.

Responsible Entity:

Whilst ultimate responsibility for ensuring completion of the requirements of these phases rests with the end-user (eg. the mining company), responsibility for conducting the activities may be delegated to vendors, contractors and others.

Activities to be Undertaken:

During the operation and maintenance phases, a configuration (engineering change) management process should be implemented for ongoing management of the EUC, the EUC control system, the safety-related systems and the "other risk reduction measures". This process, as far as possible, should follow established engineering change management processes, providing that the following essential requirements are being met.

1. The modification or retrofit should only be initiated by the issue of an authorised request, under procedures provided in the Configuration Management Plan
2. The reason for the modification should be documented as well as the hazards affected and the details of the proposed change
3. An impact analysis of the modification should be performed to establish the effect of the proposed changes on the functional safety of the EUC, EUC control system, safety-related systems or "other risk reduction measures"
4. The assessment should include a hazard and risk analysis sufficient to determine the necessary breadth and depth of subsequent activities
5. All accepted modifications that have an effect on the EUC, EUC control system, safety-related systems or "other risk reduction measures" should initiate a return to the appropriate phase
6. A complete action plan should be prepared and documented before carrying out any modifications
7. Details on the specific requirements of these configuration management activities will be included in the verification and validation plans

A model modification flowchart based on these requirements (from AS61508.1 Figure 9) is shown in Figure 7.24.

Figure 7.24: AS(IEC)61508.1 Model Modification Procedure

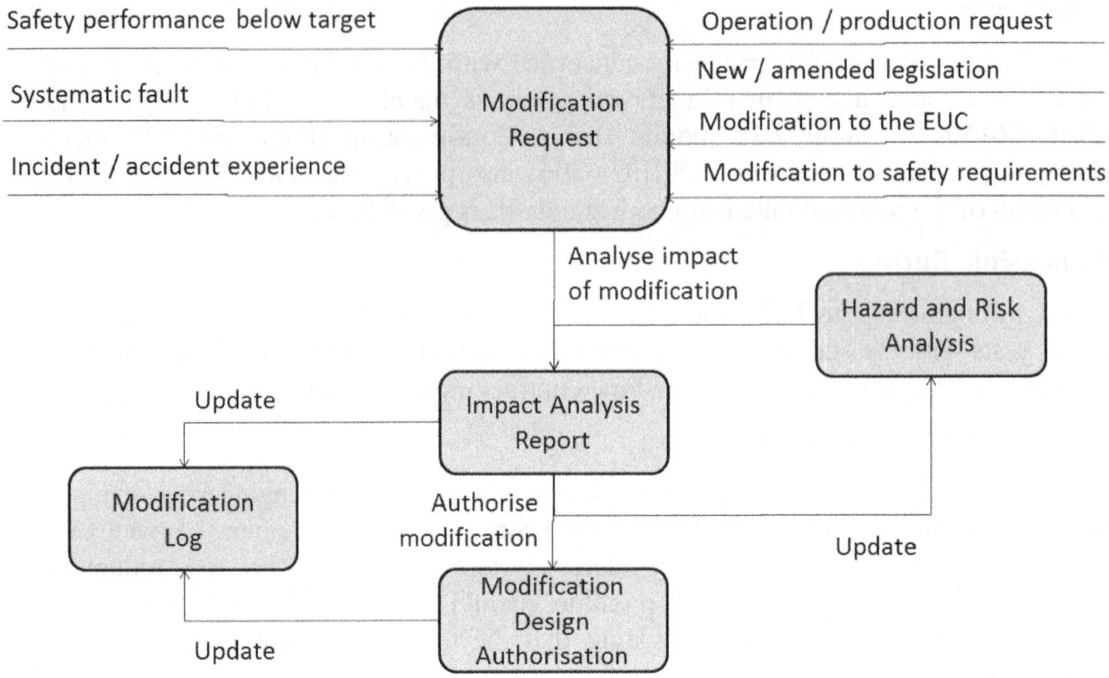

Documentation to be Kept:

The information and results acquired should be maintained throughout the overall safety lifecycle. The Safety Lifecycle Dossier should be updated with the following information:

1. A plan of the modification process
2. Documentation of the decision-making process and each decision affecting a safety-related system
3. Procedures for defining a unique baseline of each version of the safety-related systems
4. Means for definition of all configuration items of a baseline.
5. A chronological documentation (e.g. logbook) of the change request procedures including:
a. Identified hazards which can be affected
b. Description of the change request (hardware and/or software)
c. Reason(s) for the change request (see also 9.2.1)
d. Decision made (and authorization for each decision)
e. The impact analysis
f. Re-verification (of each phase) and revalidation
g. All documents affected by the change request activities
h. All activities which were carried out during the change process and the persons/entities who were responsible for them

i. Any other documents that are generated by the return to a previous safety lifecycle phase (and all subsequent phases)
6. Documentation of the following information to permit a subsequent audit:
a. Configuration status
b. Release status
c. The justification for and approval of all modifications
d. The details of the modification

7.25 AS(IEC)61508 Phase 16 - Decommissioning and Disposal

Objective/s:

AS(IEC)61508 Phase 16 is primarily concerned with ensuring that the functional safety of the safety-related systems and "other risk reduction measures" is appropriate for the circumstances during and after the activities of decommissioning or disposal of the EUC. Decommissioning and disposal may generally be treated in the same way as a modification / retrofit phase, recognising the special nature of the decommissioning / disposal activity.

Responsible Entity:

Whilst ultimate responsibility for ensuring completion of the requirements of these phases rests with the end-user (eg. the mining company), responsibility for conducting the activities may be delegated to vendors, contractors and others.

Activities to be Undertaken:

From AS(IEC)61508.1 Clause 7.17 the following should occur:

1. Prior to decommissioning / disposal, an Impact Analysis should be carried out and should include an assessment of the impact of the proposed activity on the functional safety of the EUC, EUC control system, safety-related systems or "other risk reduction measures"
2. The Impact Analysis should also consider adjacent plant and systems and their impact on the EUC, EUC control system, safety-related systems or "other risk reduction measures"
3. The assessment should include a hazard and risk analysis sufficient to determine the necessary breadth and depth of subsequent functional safety activities
4. Decommissioning / disposal should only be initiated by the issue of an authorised request, which should be dependent on the results of the Impact Analysis
5. Prior to decommissioning / disposal taking place, a plan should be prepared which should include procedures for the closing down of the EUC, EUC Control System, safety-related systems or "other risk reduction measures" and their dismantling
6. If decommissioning / disposal has an impact on the functional safety of the EUC, EUC control system, safety-related systems or "other risk reduction measures", it should initiate a return to the appropriate lifecycle phase in order to address those inadequacies identified

Verification requirements should be determined on the basis of the expected impact of the decommissioning / disposal activity on the functional safety achieved by the EUC, EUC control system, safety-related systems or "other risk reduction measures", and whether a return to earlier lifecycle steps is necessary.

Documentation to be Kept:

The information and results acquired should be maintained throughout the overall safety lifecycle. The Safety Lifecycle Dossier should be updated with the following information:

1. Details of the Impact Analysis of decommissioning / disposal upon the functional safety of the EUC, EUC control system, safety-related systems or "other risk reduction measures"
2. Decommissioning / Disposal Request
3. Decommissioning Plan
4. Any other documents that are generated by the return to a previous safety lifecycle phase (and all subsequent phases)

8. Other Issues in Functional Safety

8.1 Functional Safety of Legacy (Existing) Plant - The *FAST Functions Allocations Systems Traceability*® Methodology.

Whilst it is the intention of AS(IEC)61508 that functional safety activities commence at the conception stage of a project, little mention is made in the standards of how to approach legacy (existing) plant that was conceived, specified, designed and installed prior to the advent of AS(IEC)61508.

The issue with such plant is that it has already reached Phase 14 (Operations and Maintenance) of the AS(IEC)61508 safety lifecycle, without having had to pass the strict regime of requirements of the previous thirteen (13) phases. It is unlikely that a quantified level of safety integrity has neither been determined and specified in the early stages of design of the controls and protective functions, or verified as having been achieved at the commissioning phase.

Following several mine safety incidents and accidents, regulatory authorities suggested, via safety alerts and the like, that mines should also review the design of their legacy plant control and protective systems and upgrade their systems as appropriate via a 'functional safety approach'[49]. The use of the AS(IEC)61508, AS(IEC)62061 and AS4024.1 standards have been suggested.

It would require an extraordinary effort in terms of time, cost and manpower to revert to Phase 1 and retrospectively progress through the thirteen (13) missed phases of the safety lifecycle for every item of legacy plant at a mine. However, it is clear that ageing plant, whose controls and protection systems are unlikely to have been designed to meet rigorous targets on safe failure, fault tolerance and reliability may not be capable of achieving the levels of risk reduction that now expected of such systems. Clearly, a compromise needs to be struck.

In response to this situation I developed the ***FAST Functions Allocations Systems Traceability***® process[50]. This is a stream-lined approach to evaluating the adequacy of legacy plant control and protective systems using established machinery 'functional safety' principles.

FAST Functions Allocations Systems Traceability® is a four (4) step methodology whereby:

1. Safety-related control and protective functions are identified (**F**unctions)
2. Levels of reliability performance (ie. SIL's or PL's or CAT's) are allocated to the safety functions retrospectively (**A**llocations)
3. The capability of the systems which implement the safety functions are assessed and improved where necessary (**S**ystems)
4. A Functional Safety Dossier containing all analysis records and other documentation which is produced during the first three steps, or which is relevant in demonstrating the achieved safety integrity of the functions is created (**T**raceability)

[49] See New South Wales Department of Primary Industries Safety Alerts: SA09-03, SA06-13, SA06-12, and SA06-01.

[50] *FAST Functions Allocations Systems Traceability*® is a registered trademark of Marcus Punch Pty. Ltd.

An overview of ***FAST Functions Allocations Systems Traceability***® is provided in Figure 8.1. I developed a half-day training workshop in 2009 to introduce the process to the New South Wales mining industry and this has been well received, with more than 200 course participants being introduced to the process since late 2009.

Figure 8.1: FAST Functions Allocations Systems Traceability®

Functions

To Do:

1. Prepare the Functional Safety Management Plan (FSMP).
2. Gather Information.
   - Risk Assessments.
   - Functional Specifications.
   - Control & Elec. Drawings.
   - Incident / Accident data.
   - Other relevant data.
3. Group 'Brainstorm'.
   - All concerned parties.
   - Multi-disciplinary.
   - Include operators.

To Document

1. The control and protective functions (whether safety functions or not).
2. The general hazards being prevented / controlled by them.

Allocations

To Do:

1. If a function is a safety function, select PL allocation parameters and determine the PL required:
   C: potential injury,
   F: exposure of persons,
   P: probability of avoidance,
   W: frequency of demands.
2. Determine whether the safety function will be E/E, PE or 'other technology'. Convert the PL to a SIL or CAT as appropriate.
3. Determine the 'demand mode' (from the W parameter).

To Document

1. The parameters selected for each function.
2. The list of functions which are safety functions.
3. Implementation – E/E, PE or 'other technology'.
4. The PL / SIL / CAT allocations and 'quantitative' analysis for those safety functions requiring it.
5. The 'demand modes'.

Systems

To Do:

1. Functional Block Diagrams (FBD's) of each safety function,
2. Determine the SIL Claim Limit (SILCL) or 'designated' CAT architecture of each safety function.
3. Calculate the PFD/ PFH of each safety function, as appropriate.
4. Tabulate the achieved SIL / CAT against the allocations.
5. Check systems against the conformance checklists
6. Upgrade the design / maintenance / operations to achieve the allocated SIL / CAT where a shortfall exists.

To Document

1. FBD's, architectural analysis and calculations. (incl. test / checking / replacement intervals of all parts).
2. Table of achieved SIL / CAT against allocations.
3. Conformance template checklist.
4. Functional & Safety Requirements Specification (where upgrades required).
5. Upgraded design and operation (per the F&SRS) including design verifications.

Traceability

To Do:

1. Gather all documents produced into a logically referenced hardcopy or electronic dossier.
2. Obtain a peer review of compliance to the FAST™ process and the adequacy of the dossier contents.
3. Prepare an executive summary for the dossier:
   - Confirm that the process has been audited,
   - Confirm that all safety function shortfalls have been addressed.
4. Recommence Phase 14 of the AS61508 safety lifecycle.

To Document

1. All documents produced.
2. Peer reviewers report.
3. Executive summary.
4. Details (incl. CV's) of those involved.

8.2 Pre-emptive SIL Allocation for Machinery Designers and Vendors

A key feature of the functional safety lifecycle is that end-users conduct risk assessments on the proposed operation of plant at their site and use those risk assessments to determine the need for safety-related systems. Safety integrity requirements (ie. SIL's, PL's or CAT's) are then allocated to the safety functions performed by the safety-related systems on the basis of the risks posed. Those requirements are then specified by the end user to contractors and/or vendors via a Safety Requirements Specification (SRS).

One of the pitfalls of this project-based approach to the safety lifecycle occurs when plant and machinery designers want to develop new plant and machinery products. In this case, a large part of the design process may occur before any direction is received from end-users regarding the safety functions and SIL's, PL's or CAT's that are required for their applications.

This is a particularly challenging situation and if a pro-active approach to functional safety is not taken by the designer or vendor, one or more of the following outcomes may occur:

- The product reaches the market-place and its safety-related systems do not comply with the specific requirements of one or more of the functional safety standards
- The product reaches the market-place and its safety-related systems do not meet customer's Safety Requirements Specifications
- Significant re-design or re-validation of the product is required to ensure that compliance with the standards and/or customer's requirements can be achieved
- Inability, or difficulties, in obtaining design registration of the product with mining or other regulatory authorities in some jurisdictions
- Inability, or difficulties, in obtaining approval to sell the product in certain countries[51]
- Significant competitive disadvantage in the event that local or foreign competitors have applied AS(IEC)61508, AS(IEC)62061, AS(IEC61511, ISO13849, AS4024.1 or associated IEC / ISO / EN standards to the design of their competing products[52]

I suggest that plant designers and vendors should try to alleviate this problem by pre-empting their potential customer's Safety Requirements Specifications (SRS's).

---

[51] For instance, since December 2009, all European Union (EU) countries have been legally obliged to conform to Machinery Directive (2006/42/EC). Manufacturers must demonstrate compliance with this directive via compliance with a set of harmonised European standards (including IEC61508 and ISO13849). Machinery that does not meet the requirements of the directive cannot be sold in the EU countries, and violations can lead to expensive delays in placing the machinery on the market or enforced recalls (see http://www.internationallawoffice.com/Newsletters/detail.aspx?g=d2a7ec89-fd1b-43ea-b8aa-fc46fec4cbcf for details).

[52] I suspect that this issue may eventually drive Australian manufacturers of new mining equipment and safety products to adopt AS(IEC)61508 whole-heartedly during the design process. There is clear evidence that IEC61508 and associated standards are being widely adopted by manufacturer's worldwide. For instance, by November 2002, the Chinese standards agency had adopted 1,911 or 45.8% of IEC International Standards and had plans to eventually adopt all remaining IEC standards (see http://www.iec.ch/news_centre/release/nr2002/nr1402.htm for details).

By performing a preliminary SIL / PL allocation for themselves using a naturally conservative SIL allocation methodology, the product designer / vendor will get a reasonable idea of the safety-related systems and SIL or PL requirements that may eventually be specified by their customers. This will enable the product-development process to proceed with a sufficient degree of confidence that the product and its safety-related systems will be able to comply with the requirements of most customers applications.

Indeed, this pre-emptive safety lifecycle work should be viewed as a source of competitive advantage – which can be used during marketing and tendering to demonstrate the established functional safety capabilities of the vendor's products. The pre-emptive SIL allocations will also provide a clear design baseline for further product development.

Eventually when customers place orders for the product, much of the safety lifecycle work for the 'realisation' stage of the safety-related systems will have been done and can be promptly delivered. Moreover, if a conservative safety requirements allocation methodology is used to guide the specification and design of the safety-related systems, it is far less likely that they will require re-design at a later date because they do not meet the requirements of a particular customer's SRS.

The general scheme for vendor pre-emptive safety requirements allocation is shown in Figure 8.2.

Figure 8.2: Vendor Pre-emptive SIL / PL Allocation

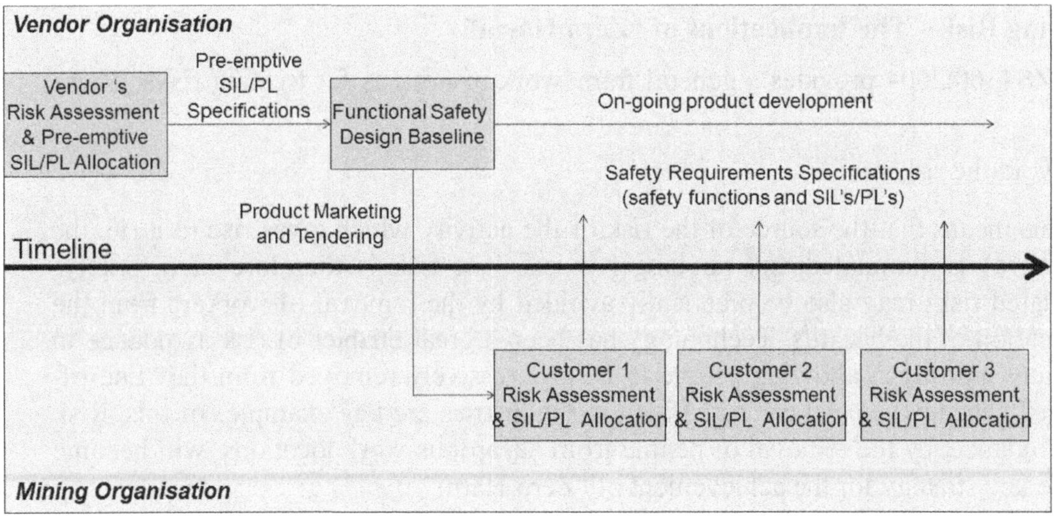

It must be stated, however, that if a designer / vendor follows this approach during their product's 'realisation' it provides no justification for their customers (ie. the mining companies) to forego their own hazard and risk analysis and SIL / PL allocation activities.

If the intent of AS(IEC)61508 is to be met, the first five phases of AS(IEC)61508 must be performed by the end-user (ie. the mining company). This approach ensures that a confirmatory Safety Requirements Specification is developed for the end-user's specific application of the product. Then, the safety requirements for the specific application can be compared with the safety requirements already designed into the product. This approach should allow the end-user to evaluate whether the product, as designed, has a sufficient level of functional safety for their specific intended use.

8.3 Reconciling Functional Safety / System Safety with "Zero Harm"

Author's Note:

This section of the book has been adapted from a paper given by the author to Australian mining, hazardous area and risk engineering conferences during 2009 and 2010. Additional emphasis on the functional safety aspects has been included here.

Defining 'Zero Harm'

It is important that the key concepts of risk management are understood, for the "Zero Harm" approach to be understood.

AS/NZS4360:2004 provides a list of standard risk management definitions, including:

Hazard: A source of potential harm.[53]

Unfortunately, "harm" has not been defined in the standards. We must therefore turn to accepted definitions of the word, which include - injury, hurt, damage, detriment, misfortune or loss.

Risk: The chance of something happening that will have an impact on objectives.

Safety: Freedom from harm.[54]

"Zero Harm" is therefore a term that describes the absolute achievement of "safety". A "Zero Harm" workplace is one where the potential for harm to workers has been removed absolutely.

Treating Risk – The Implications of "Zero Harm"

AS/NZS4360:2004 provides a general framework of options for treating risks. These are:

- Avoid the risk

  This means that the source of the risk or the activity which gives rise to it (ie. the hazard) is eliminated and no longer exists. The risk is therefore zero. Safety-related risks may also be practically avoided by the removal of workers from the location of the hazards. Technology has been a great enabler of risk avoidance in many industries, allowing people to be progressively removed from the "line-of-fire". The textiles and car manufacturing industries are key examples of this. Risk avoidance, by the removal of people from hazardous work locations, will become the key strategy for the achievement of "Zero Harm".

  A current example of the risk avoidance strategy is mine automation. Here, persons are removed from direct exposure to operating plant and equipment. High integrity electronic systems are provided to prevent worker access into autonomous operating zones of the mine and for plant isolation during maintenance or other worker access. For instance, the Australian Centre for Field Robotics (ACFR) at the University of Sydney has established a major new Centre for Mine Automation. Rio Tinto, a global mining company, has committed $21m of funding for an initial period of five years for this new centre. The aim of the Rio Tinto Centre for Mine Automation (RTCMA) is to develop and implement the vision of a fully autonomous, remotely-operated mine. Automation and field

---

53 This definition has been sourced from ISO/IEC Guide 51.

54 This definition has been sourced from ISO/IEC Guide 51.

robotics will play an increasingly important role in mining and other hazardous industries in the future.

- Reduce the risk

  This means that the likelihood and/or consequences of the risk are reduced. Historically this has been the key option for treatment of risk, when means have not been available to avoid risk. The "Zero Harm" approach in its strictest sense does not tolerate any risk which is not zero, or has not been absolutely eliminated. In practice however, "Zero Harm" must be seen to be a long-term incremental process. Whilst means may not exist to eliminate every hazard now, they may exist in the future. Between now and that time, a practical interim approach would be one that allows incremental, measurable improvement of non-zero risks.

- Share the risk

  This option was described in previous versions of AS/NZS4360 as 'risk transfer'. Certain aspects of risk may be shared or transferred, and this is the basis of the insurance industry. The financial consequences of industrial accidents or other events may be mitigated by insurance, but physical harm to workers cannot. Furthermore, Australian federal and state OH&S legislation does not allow liabilities for safety to be delegated to others. The use of insurance implies that risk has not been eliminated. If there was no risk, insurance would not be necessary.

- Retain the risk

  This option was described in previous versions of AS/NZS4360 as 'risk acceptance' and is related to the principle of As Low As Reasonably Practicable (ALARP). According to this principle, the residual risk of a hazard may be tolerable if there are no further reasonably practicable means available to reduce it further. Australian federal and state OH&S legislation requires that if a hazard cannot be eliminated, then the risk posed must be reduced to ALARP. "Zero Harm" implies that ultimately the only acceptable risk is one that does not exist, because it has been eliminated or otherwise reduced to zero.

Towards Zero Harm – A Reality Check

"Zero Harm" implies absolute safety. Ultimately, risks must be avoided - by eliminating the sources of harm or removing people from them. "Zero Harm" is a noble vision, but one that requires a modicum of common sense applied to it in order for it to have any chance of success. AS/NZS4360:2004 offers sage advice on this issue:

Risk avoidance can occur inappropriately if individuals or organisations are unnecessarily risk-averse. Inappropriate risk avoidance may increase the significance of other risks or may lead to the loss of opportunities for gain.

(from AS/NZS4360:2004 Section 3.6.3)

In practice, "Zero Harm" must be seen as a "vision" rather than a "goal", as it is often called by its proponents. Practically, "Zero Harm" will need to be a long-term process, based on incremental risk reduction and gradual exploitation of technologies for risk avoidance. Whilst means may not exist to eliminate every hazard now, they may exist in the future. Between now and that time, a practical interim approach would be one that allows incremental, measurable improvement of non-zero risks.

Towards Zero Harm – Where are we now?

In recent decades many advances and achievements have occurred in the field of workplace safety. In fact, advances in safety have occurred incrementally across many centuries. We now live in an era of very low workplace fatality and injury rates. A few examples from history demonstrate the extent of the changes:

- The 6700 km long Great Wall of China was constructed during 5000 to 200BC, with the loss of 2 – 3 million workers[55]. The safety statistics for this project were as follows:

  Safety Performance: 500 fatalities per year

  400 fatalities per constructed kilometre of wall

- In contrast, Australia's Snowy Mountain Scheme which was constructed from 1949 to 1974, resulted in the loss of 121 workers[56].

  Safety Performance: 5 fatalities per year

  1 fatality per constructed kilometre of tunnel

Statistics from the Australian mining industry[57] also reveal some interesting results in safety performance during the last decade or so.

The rate of occurrence of Lost Time Injuries (LTI's) on a "per million man-hours" basis has significantly and consistently decreased between 1994 and 2009. Figure 8.3 shows the trend, which has been consistently downward. Whilst the rate of decrease in the LTI Frequency Rate (LTIFR) is reducing, an extrapolation of this highly correlated year-to-year data would indicate that the LTIFR may be lower than four (4) LTI's per million man-hours by 2010, compared to 25 LTI's per million man-hours recorded in 1994. Clearly, advances in OH&S legislation, workplace Safety Management Systems, organisational cultural change and other areas have had a significant positive impact for worker safety.

---

55 http://en.wikipedia.org/wiki/Great_Wall_of_China

56 Fergus Robinson , "A study of occupational health and safety practice in the construction of the Snowy Mountains Hydro-Electric Scheme", National Occupational Health and Safety Commission

57 Minerals Council of Australia, Safety & Health Performance Report of the Australian Minerals Industry 2006-2007.

Figure 8.3: Australian Mining Lost Time Injury Frequency Rate (LTIFR) 1994-95 to 2006-07

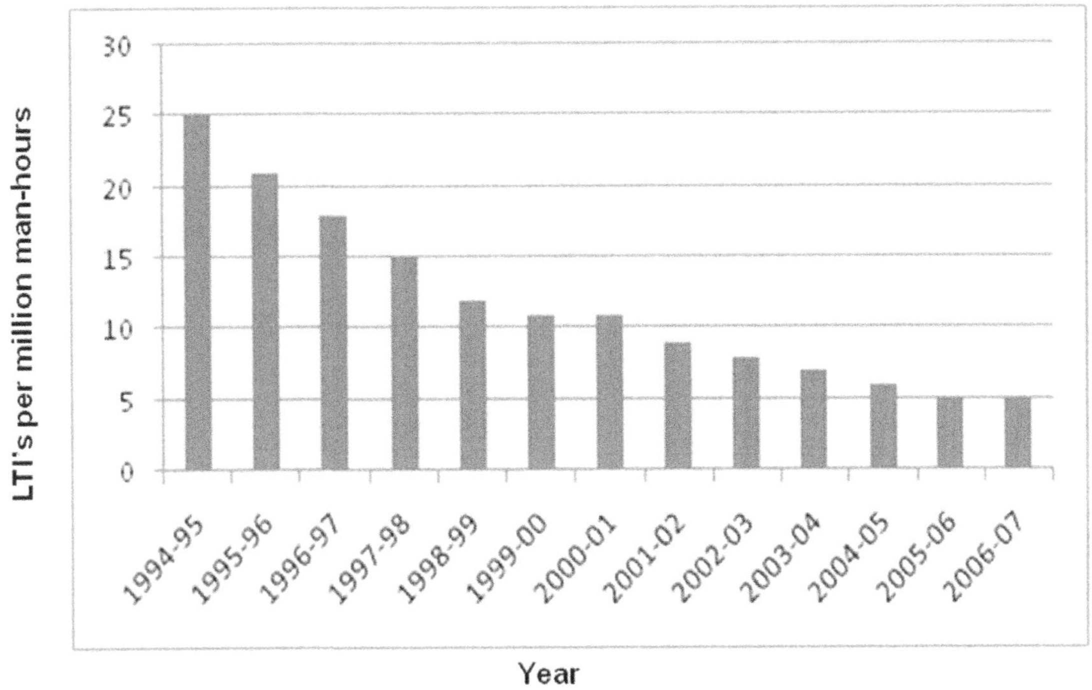

The corresponding Fatal Injury Frequency Rates (FIFR) for the Australian mining industry tells another story. The average FIFR has decreased from around 0.15 in 1994 to around 0.05 in 2009, however there are wide variations in the annual fatality rate figures. Figure 8.4 shows the trend. The year-to-year data is poorly correlated and it would be very difficult to predict with any confidence what the FIFR might be in the future, or even if it is likely to continue to decrease or stabilise. Whilst there is other anecdotal evidence to support a case that advances in OH&S legislation, workplace Safety Management Systems, organisational cultural change and other areas have had a positive impact on worker fatality rates, the year to year performance has been inconsistent and subject to significant variations.

The conclusion to be drawn is that the Australian mining industry has had, and continues to have, significantly less control over the incidence of workplace fatalities than it does for LTI's.

Figure 8.4: Australian Mining Fatal Injury Frequency Rate (FIFR) 1994-95 to 2006-07

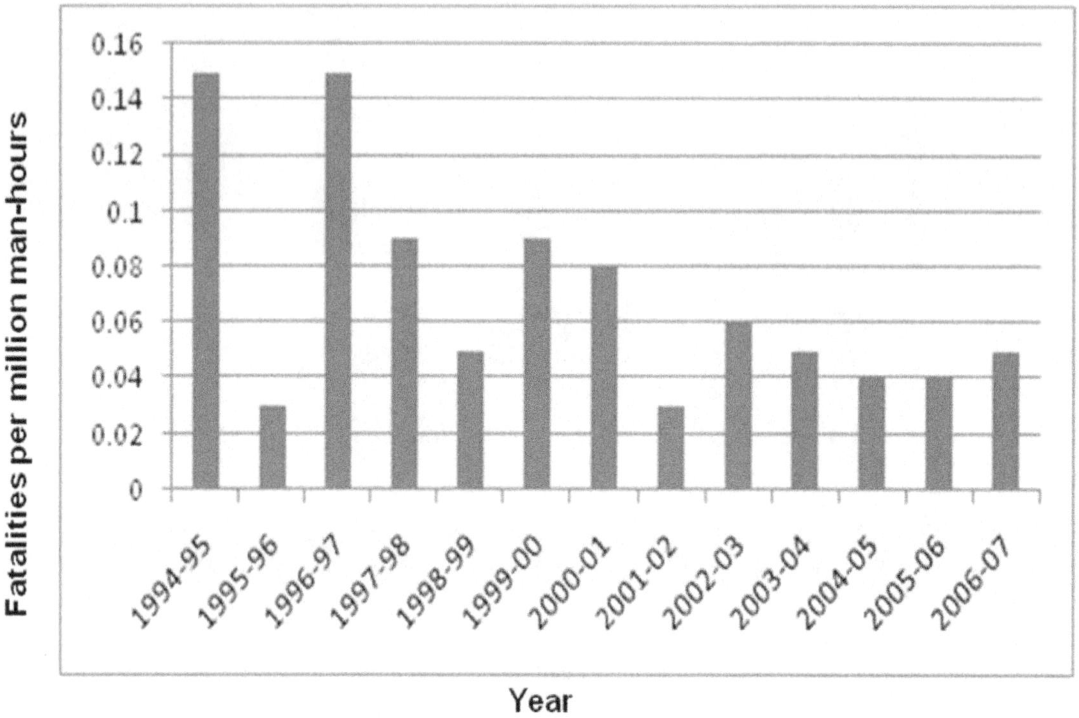

Fatality rates for other industries and activities have also been calculated[58], and a summary is provided below in Figure 8.5. The Australian mining industry figures have been converted from FIFR (measured in "per million man-hour") to an Individual Risk (of death) Per Annum (IRPA)[59] in order to provide a basis for comparison.

[58] The primary reference was BHP Billiton HSEC Toolkit No. T06 Rev.2 "Quantitative Risk Assessment Approach", pages 14-15. This document references a number of data sources including: Lees, *Loss Prevention in the Process Industries* - (Ed 2) App 14 (all equipment) ISBN 0 7506 1547 8, UK HSE : MKIDAS Accident Database and UK IChemE Accident Database and others. Figures for the Australian mining industry and Australian general industry have been inserted on the basis of figures provided by the Minerals Council of Australia Safety Performance Reports for 2003-04 and 2006-07 and the Minerals Industry Safety and Health Centre (MISHC) Final Report to the Mineral Resources Council on Underlying Causes of Fatalities and Significant Injuries in the Australian Mining Industry, 4th March 2005 (Appendix 1, page 2).

[59] Assuming that the average worker performs 2000 hours of work per year.

Figure 8.5: Comparative IRPA's for Selected Industries / Activities.

Industry / Activity	**Individual Risk Per Annum (IRPA)**
Cigarette Smoking (20 / day)	0.0050
UK Sea Fishing	0.0012
UK Oil & Gas	0.0010 (including 167 Piper Alpha fatalities)
Australia All Mining (1993-1994)	***0.0003***
UK Coal Mining (1987-1991)	0.0002
Australia All Mining (2006-2007)	***0.0001***
Soccer / Rock Climbing	0.000040
Australia All Industry (1996-1997)	***0.000038***
UK Energy Production	0.000025
Australia All Industry (2001-2002)	***0.000024***
Contraceptive Pill	0.000020
UK All Industries (1987-1991)	0.000018
Car Driving	0.000017
UK Education / Teaching	0.0000015
UK Nuclear Accident / Lightning	0.00000010
Aircraft Crash	0.00000002
Struck by Meteorite	0.00000000006

The conclusion to be drawn is that the Australian mining industry has had significantly less control over the incidence of workplace fatalities than general Australian industry, but the gap is closing.

Covering the Risk Spectrum

The Australian mining industry LTIFR and FIFR results may be explained as follows.

Within any facility there exists a wide range of hazards. The risks posed by these hazards may be generally characterised and assessed by their consequences, likelihood of occurrence and exposure of persons to them. Some hazards will be high likelihood / low consequence – such as minor plant breakdowns, LTI's and minor environmental releases. Other hazards will be less likely but with more significant or even catastrophic consequences including fatalities to workers and/or the public and serious or catastrophic damage to the facility or local environment. Figure 8.6 provides a graphical depiction of the risk spectrum.

Figure 8.6: The Risk Spectrum.

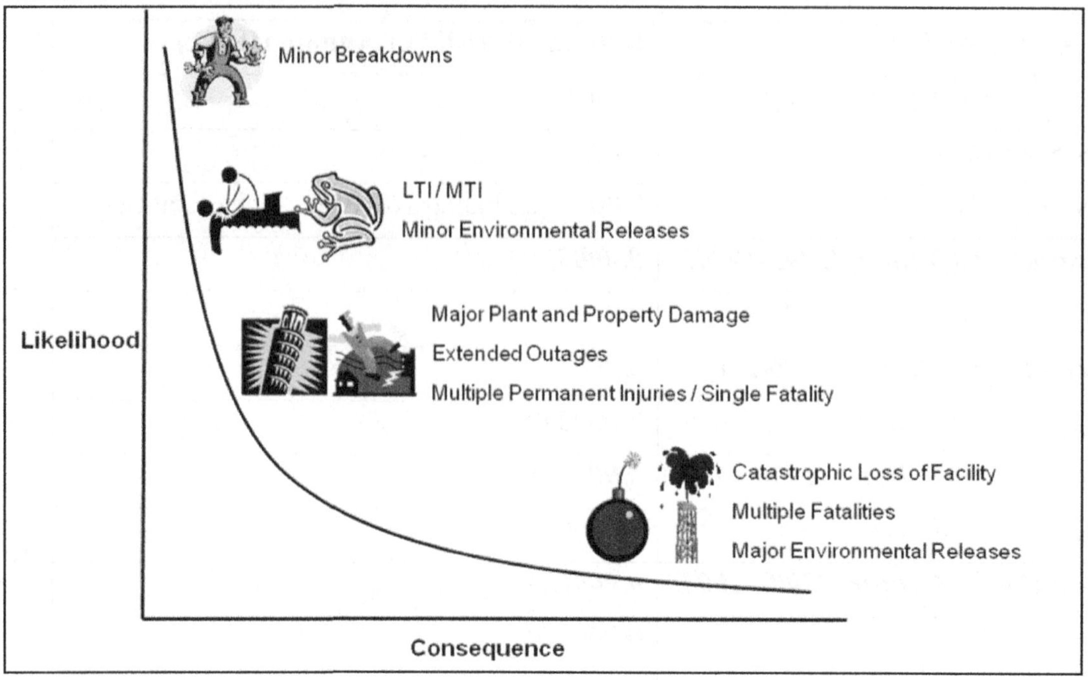

Clearly, the mining industry has achieved a good level of control of the high likelihood / low consequence hazards (shaded area) as a result of advances in OH&S legislation, workplace safety management systems and other measures introduced to reduce the incidence of injuries sustained in the course of performing workplace tasks.

The mining industry has not yet gained control over the low likelihood / high consequence events. Such hazards are more likely to be a result of process hazards or technical failures in plant, machinery or systems (eg. unplanned movements, fire / explosion, toxic release) rather than hazards related to workplace tasks. The conclusions drawn from recent industrial disasters indicate that this is a common failing across many industries. The following quotes illustrate this point:

"Safety rather than engineering has dominated the lessons from the disaster. The safety focus has resulted in risk management being aimed at reducing high-frequency, low consequence personal injuries. This is different from an engineering focus that is essential to reducing low-frequency, high-consequence catastrophes. A balance needs to be struck that does not neglect either end of the risk spectrum".

("Have Australia's Major Hazard Facilities Learnt from the Longford Disaster?", IEAust, October 2001.)

"Their view is that our system was insufficient to systematically and continuously identify, reduce and manage risk, in particular to identify low probability high consequence risks."

(Lord Browne, CEO of BP plc. making remarks at a press conference on 16th January 2007 in response to the release of the Baker Panel Report into the Texas City refinery accident.)

Clearly, to achieve the "Zero Harm" vision, a technical strategy for reduction and avoidance of the high consequence / low likelihood hazards is required.

Incremental Risk Reduction – The System Safety Approach

The "System Safety" approach provides a framework for incremental, measurable improvement of non-zero risks. System Safety engineering is about establishing and achieving the requirements for, and the characteristics, intended behaviour and integrity (reliability) of safety-critical systems and other safeguards for plant and equipment via a risk management process.

Integral to the System Safety approach is the definition of "safety targets". There are numerical criteria used as a means to measure, evaluate and improve safety performance of systems and plant. Safety targets have already been suggested by various organisations and industry groups throughout the world. The United Kingdom Health & Safety Executive (UK H&SE) and the United Kingdom Institution of Gas Engineers and Managers (IGEM) are two groups who have been at the forefront of these developments. These two bodies have suggested that the Individual Risk (of death) Per Annum (IRPA) to a worker should be no more than 1 in 1,000 (or 10^{-3}) considering all workplace hazards they are potentially exposed to. They have also suggested that an IRPA of less than 1 in 1,000,000 (or 10^{-6}) could be considered to be "negligible" or "broadly acceptable".

In Australia, safety targets have also been considered for the mining industry:

"Safe" industry should be about 0.005 fatalities / million man-hours, or a factor of ten lower than currently reported in the mining industry.

(Final Report to Queensland Resources Council (QRC) on underlying Causes of Fatalities and Significant Injuries in the Australian Mining Industry, Minerals Industry Safety and Health Centre (MISHC), March 2005, page 11).

The current Australian mining industry Fatal Injury Frequency Rate (FIFR) is approximately 0.05 fatalities per million man-hours. For an individual worker, performing on average 2,000 hours of work per year, this figure translates to an average Individual Risk (of death) Per Annum (IRPA) of about 1 in 10,000 (or 10^{-4}) across the mining workforce. The MISHC report suggests that a "safe" industry FIFR target should be 0.005 fatalities per million man-hours, or one tenth of the current FIFR. This figure translates to an average IRPA of 1 in 100,000 (or 10^{-5}) across the Australian mining workforce. Therefore, the FIFR currently achieved is ten times the suggested FIFR target.

The concept of these targets is shown in Figure 8.7.

Figure 8.7: Overall Risk Targets

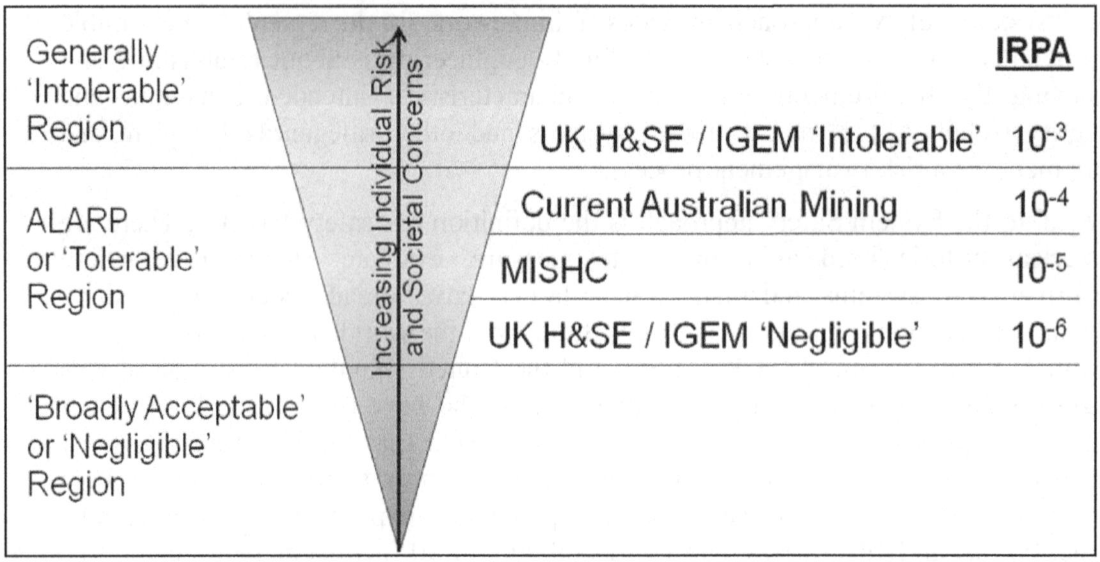

The UK H&SE and IGEM suggest that a "broadly acceptable" risk level requires an IRPA of less than 1 in 1,000,000 (or 10^{-6}). This figure translates to a FIFR of less than 0.0005 fatalities per million man-hours, which is one tenth of the above MISHC "safe" industry FIFR target, and one hundredth of the current FIFR performance. Clearly, safety performance needs to significantly improve if the 'MISHC' and 'Negligible' IRPA targets are to be achieved.

Towards Quantitative Risk Analysis (QRA)

It has been seen that safety performance in the high consequence / low likelihood part of the risk spectrum is deficient. The "Zero Harm" technical strategy that I have suggested may require organisations to set and measure incremental numerical safety targets. It is entirely appropriate that high consequence / low likelihood hazards and the risk controls that are used to reduce them are analysed on a quantitative basis. QRA is already used in industries where the System Safety approach is mature, such as railways, oil / gas and defence / aerospace.

These risk targets expressed previously have been in terms of the overall risk to a worker considering all hazards. However, in practice risks are identified and assessed on an individual hazard basis and consider all workers who are potentially exposed. For the purposes of QRA the IRPA safety targets may need to be translated to a 'per hazard' basis. A scheme for doing this is provided in Table 8.1which has already been described in Section 7.5. The bolded figures are explained below.

Table 8.1: "Per Hazard" Safety Targets

IRPA Target (fatalities / employee / year)	No. of Employees at Site	Site-wide Fatality Target (fatalities / year)	No. of Fatal Hazards Present	Per Hazard Frequency Target (events / year)
			10	10^{-6}
	1	10^{-5}	100	10^{-7}
			1000	10^{-8}
			10	10^{-5}
	10	10^{-4}	100	10^{-6}
			1000	10^{-7}
			10	10^{-4}
10^{-5}	**100**	**10^{-3}**	**100**	**10^{-5}**
			1000	**10^{-6}**
			10	10^{-3}
	1000	**10^{-2}**	**100**	**10^{-4}**
			1000	**10^{-5}**
			10	10^{-2}
	10000	10^{-1}	100	10^{-3}
			1000	10^{-4}

The 'per hazard' safety target takes account of the chosen IRPA target, the size of the workforce and the number of hazards posed to the employees. For a typical mine site with 100 to 1000 employees continuously present (bolded), and with 100 to1000 fatal hazards (bolded) posed to each employee the suggested hazard frequency target is 10^{-4} to 10^{-6} per year (bolded). A nominal target of 10^{-5} per year is suggested.

This target may be adjusted for those hazards which could result in multiple fatalities, or in less serious safety consequences. For example, a hazard with the potential for ten (10) fatalities could be assigned the target of 10^{-6} per year, one-tenth of the single fatality target. A hazard with the potential for serious injury might be assigned a target somewhere in the range of 10^{-4} to 10^{-5} per year.

The initial emphasis of this approach should be that the high consequence events are properly identified (via a HAZOP, Broad Brush Risk Assessment or other means) and their risk controls identified. Then, QRA techniques such as Fault Tree Analysis (FTA), Event Tree Analysis (ETA), Cause-Consequence Analysis (CCA), Layer of Protection Analysis (LOPA) and others should be used to demonstrate that the proposed risk reduction strategy for the hazard will enable the nominated safety target for the hazard consequences to be achieved. An example of QRA applied to hazardous area safety is provided in Figure 8.8 using the Layer of Protection Analysis (LOPA) technique. Clearly, software tools and risk engineering skills will be needed to facilitate and record this process – but these tools and skills are already present in industry.

Figure 8.8: Hazardous Area QRA Example Using LOPA

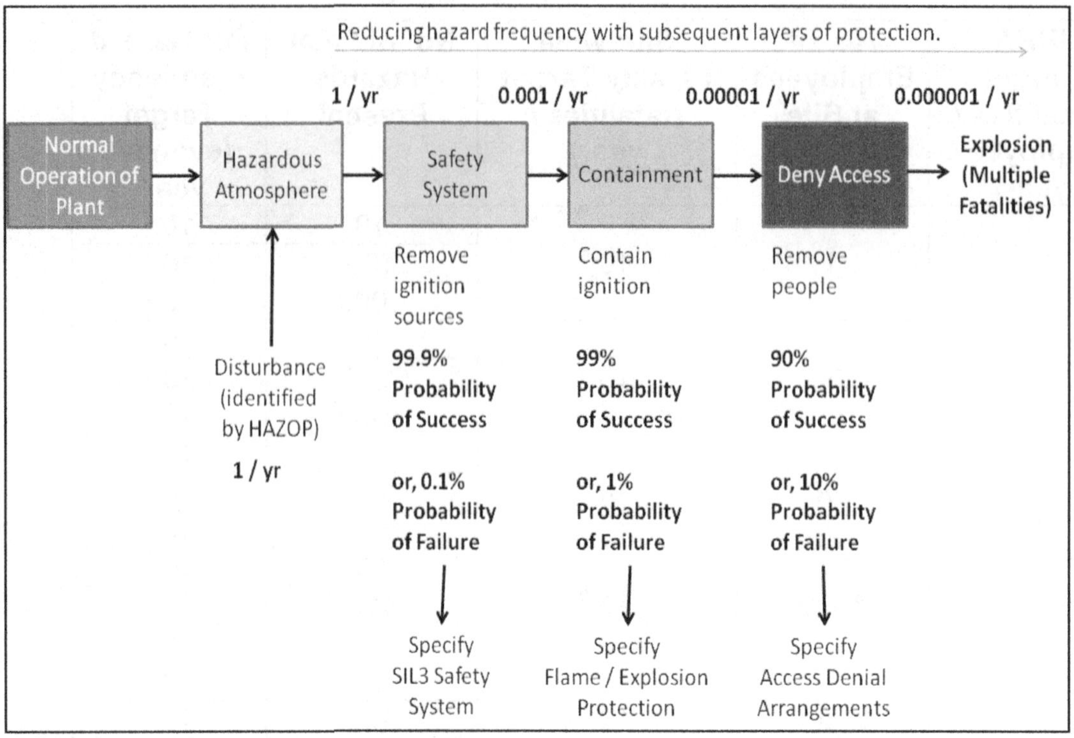

Human Factors

The effect of human error in major industrial accidents cannot be under-estimated. MISHC has conducted research into 109 accident cases and found in 107 (ie. 98.1%) that an "unsafe" human act or error was a contributor[60]. The findings were summarized as follows:

- 57 % unintentional mistakes made
- 20% cultural violations (organisationally tolerated intentional violations of rules or procedures)
- 17 % unintentional slips or lapses
- 6 % intentional or deviant violations of rules or procedures

The potential for human error or unsafe acts to contribute to accidents cannot be understated.

In this light, the risk reduction stage of "Zero Harm" should also be about substituting high-integrity engineering safeguards for those based on human actions. Practically though, for some time yet there will be situations where human actions, such as conformance to procedures, rules and emergency actions are required to prevent or control high consequence hazards. In these circumstances, human error probabilities should be established within the QRA in order to calculate the risk posed and evaluate whether the prevailing safety targets can be met. Techniques such as THERP (Technique for Human Reliability Prediction) and HEART (Human Error Analysis and Review Technique) are already available and being used to evaluate human error probabilities in some industries.

[60] MIHSC Fatal Injuries Causal Analysis (a sub-project of a research project for the Queensland Resources Council 2004/2005), pages 13-17.

Towards Zero Harm – A Framework for Improvement

It has already been suggested that safety targets will need to be redefined from time to time in order to incrementally approach the "Zero Harm" vision. It may not be practicable to change safety targets regularly, and considering the duration of time it may take to realise the technologies to enable risk avoidance, I would boldly suggest a "decade per decade" system.

That is, if an IRPA of 1 in 10,000 is achieved now in 2009, then the IRPA target for 2010-2019 would be 1 in 100,000 (in line with the MISHC recommendation). For 2020-2029 the IRPA target could be lowered to 1 in 1,000,000, in line with the UK H&SE / IGEM targets for "negligible" / "broadly acceptable" risk. On this "decade-per-decade" basis, by 2040 industry would be targeting an IRPA of 1 in 10,000,000, which equates to a FIFR of 0.05 fatalities per billion man-hours. Such a level of safety roughly translates to 0.01 fatalities per year per 100,000 workers[61], rather than 10 fatalities per year per 100,000 workers as currently experienced.

When safety targets are redefined, the QRA may be revisited in order to identify measures that can be taken to reduce the frequency or consequences of the hazards, or improve the integrity of the risk controls – to allow the new target to be achieved.

This means that "Zero Harm" will need to become a part of the "continuous improvement" framework of organisations. The ISO9001 approach to Quality Management Systems is ideally suited, well used and understood by industry and is mature.

A concept diagram of the ISO9001 continuous improvement process adapted for System Safety is provided below in Figure 8.9.

Figure 8.9: "Zero Harm" and System Safety Continuous Improvement.

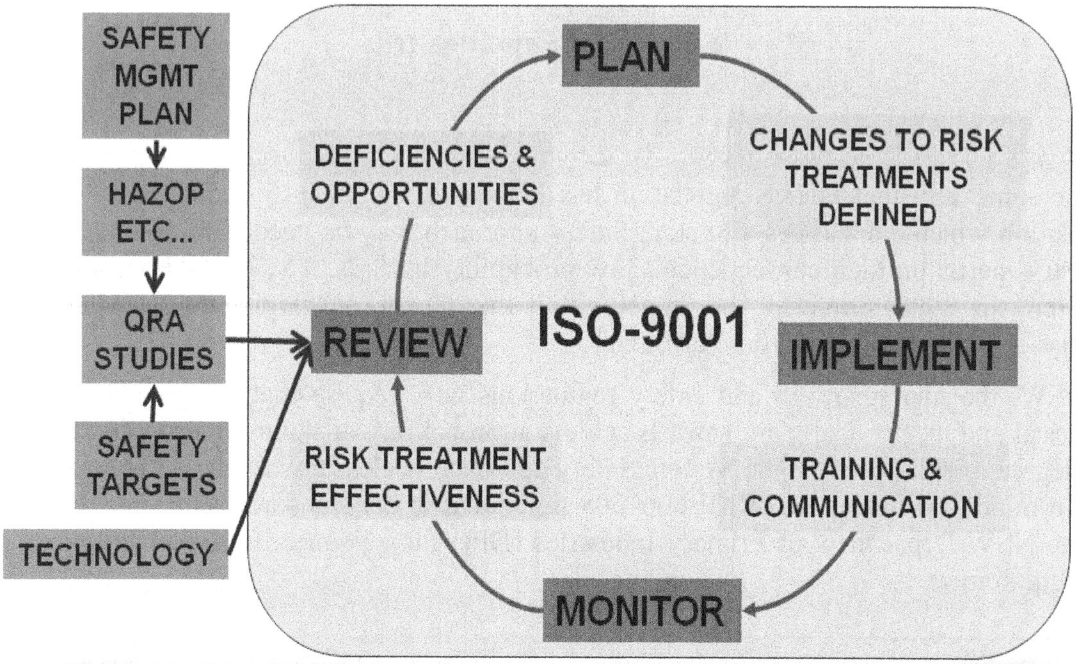

[61] Assuming 2000 work hours per person per year.

The focus should be continuous improvement of the risk spectrum. If all high consequence risks pertaining to a site are collated into a risk spectrum, this may be used as a basis for ongoing evaluation of the QRA against the overall risk targets.

In Safety Engineering the risk spectrum may be represented by an "F-N Curve", showing frequencies (F) and the number of fatalities (N) for a collated set of site hazard analyses. Such an approach has already been conceived in the general guidelines for Major Hazard Facilities (MHF's) in NSW[62]. The F-N Curve example given in that document is shown in Figure 8.10. In this graph, a predicted quantitative improvement in indicated by the 'Future Case' curve being located under and to the left of the 'Existing Case' curve.

Figure 8.10: Measuring Continuous Improvement via F-N Curves

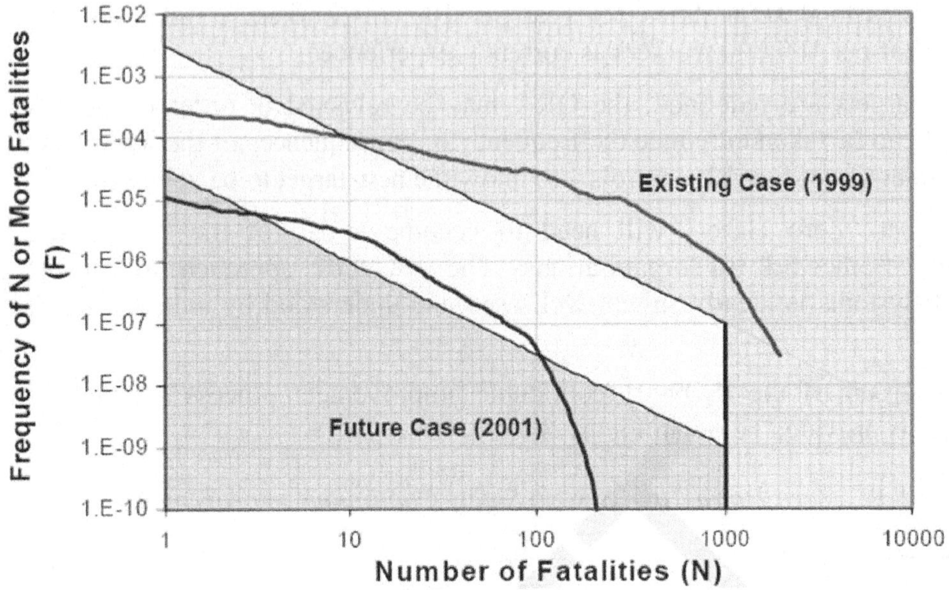

Law

In the same way that OH&S legislation has helped to significantly reduce LTIFR, legislation which emphasises a System Safety approach may be needed to overcome industry inertia on high consequence / low probability hazards. The Esso "Longford disaster" in 1998 provided the impetus for the recent developments in MHF legislation and guidelines across Australia[63].

In NSW, the mining health and safety regulations now require that safeguards for electrical and mechanical plant hazards achieve a probability of failure appropriate to the degree of risk posed[64]. The System Safety approach is clearly what the authorities had in mind, and indeed the AS(IEC)61508 functional safety standard has been cited by the NSW Department of Primary Industries (DPI) for guidance on implementing this requirement[65].

---

[62] NSW Department of Planning, Hazardous Industry Planning Advisory Paper No.3, Risk Assessment, Consultation Draft, July 2008, pages 4-6.

[63] Workcover NSW, Guide to the New South Wales Occupational Health and Safety Regulation Amendment (Major Hazard Facilities) 2008, page 5.

[64] NSW Coal Mine Health and Safety Regulation 2006, Clauses 13(1)(e)(v) and 13(1)(f)(viii).

[65] NSW Department of Primary Industries Legislation Update No.5 / 2007.

The latest draft of the federal plant regulations recommends a similar approach to the design of plant control systems[66]. A few more legislative nudges in right direction would be helpful for the "Zero Harm' cause.

Culture and Leadership

"Zero Harm" implies a significant change to the way that safety and risk management is currently practiced in Australia. The key elements will be:

- "Selling" the concept to workers
- Continued efforts to reduce the rate of cultural violations to rules and procedures
- Training of practitioners who can implement the System Safety techniques and methodologies
- Evaluation, procurement and embedding of appropriate risk engineering tools

Organisations will need to define safety targets and implement the management systems needed to measure, review and continuously improve performance.

If poorly implemented, "Zero Harm" could cost jobs, because ultimately, it will be about risk avoidance. Well implemented, it should not. For example, mine automation is expected to reduce the number of workers who directly operate mining machinery, but it will increase the number of remote control room operators and technology workers responsible for its engineering and maintenance. I expect that the pathway to "Zero Harm" will not be without its industrial challenges, and pro-active organisation leadership in areas such as skills re-training will be a key factor for a successful outcome.

Conclusion - The Enablers of "Zero Harm"

I expect the following to be important enablers for the success of the " Zero Harm" vision. These are not arranged in any sequence or order of importance.

1. Integrating a System Safety approach into the workplace risk management system
2. Exploitation of new technologies for risk avoidance (eg. mine automation)
3. Incrementally reducing risk targets (eg. "decade-for-decade")
4. Quantitative risk analysis of high consequence hazards
5. Substitute engineering safeguards for human-based ones
6. Assess and control human error and cultural violation probabilities
7. Pro-active organisational leadership and culture change
8. Whole of life-cycle effort (eg. AS(IEC)61508 / AS(IEC)61511 / AS(IEC)62061 safety lifecycle)
9. Continuous improvement (eg. ISO9001) of the risk spectrum
10. Legislative "push"

---

[66] Australian Government, Australian Safety and Compensation Council, Public Discussion Paper – Safety Requirements for the Design, Manufacture and Conformity Assessment of Plant, February 2009, page 12.

Appendices

Appendix A Glossary of Terms.

Appendix B Project Safety Lifecycle Checklist.

Appendix A: Glossary of Terms

Note 1: AS(IEC)61508.4, AS(IEC)62061 Clause 3, AS(IEC)61511.1 Clause 3, ISO13849-1 Clause 3 and AS4024.1101 provide a full complement of terms, definitions and notes.

Note 2: Those abbreviations and acronyms which are contained in each of the standards are indicated.

Note 3: In certain cases, the same terms are used by more than one standard, but with different definitions. Multiple definitions have been provided for the purpose of comparison between standards.

access time (time for access to a danger zone) (AS4024.1 term) - the time taken to access the hazardous machine parts after initiation of the stop command by the interlocking device, as calculated on the basis of an approach speed the value of which may be chosen, for each particular case, taking into account the parameters given in ISO 13855.

actuator (AS4024.1 term) - the part of the actuating system to which an external actuating force is applied.

adjustable guard (AS4024.1 term) - fixed or movable guard which is adjustable as a whole or which incorporates adjustable part(s). The adjustment remains fixed during a particular operation.

animation (AS(IEC)61508 term) - simulated operation of the software system (or of some significant portion of the system) to display significant aspects of the behaviour of the system, for instance applied to a requirements specification in an appropriate format or an appropriate high-level representation of the system design.

application (AS(IEC)61508 term) - task related to the EUC rather than to the E/E/PE system.

application software (AS(IEC)61511 term) - software specific to the user application. In general, it contains logic sequences, permissives, limits and expressions that control the appropriate input, output, calculations, decisions necessary to meet the safety instrumented functional requirements.

application software (AS(IEC)62061 term) - software specific to the application, that is implemented by the designer of the SRECS, generally containing logic sequences, limits and expressions that control the appropriate input, output, calculations, and decisions necessary to meet the SRECS functional requirements.

application software (ISO13849 term) - software specific to the application, implemented by the machine manufacturer, and generally containing logic sequences, limits and expressions that control the appropriate inputs, outputs, calculations and decisions necessary to meet the SRP/CS requirements.

application software / application data / configuration data (AS(IEC)61508 term) - part of the software of a programmable electronic system that specifies the functions that perform a task related to the EUC rather than the functioning of, and services provided by the programmable device itself.

application specific integrated circuit (ASIC) (AS(IEC)61508 term) - integrated circuit designed and manufactured for specific function, where its functionality is defined by the product developer.

architectural constraint (AS(IEC)62061 term) - set of architectural requirements that limit the SIL that can be claimed for a subsystem.

architecture (AS(IEC)61508 term) - specific configuration of hardware and software elements in a system.

architecture (AS(IEC)61511 term) - arrangement of hardware and/or software elements in a system, for example, (1) arrangement of safety instrumented system (SIS) subsystems; (2) internal structure of an SIS subsystem; (3) arrangement of software programs.

architecture (AS(IEC)62061 term) - specific configuration of hardware and software elements in a SRECS.

assessor (AS(IEC)61508 term) - person, persons or organization that performs the functional safety assessment in order to arrive at a judgement on the functional safety achieved by the E/E/PE safety-related systems and other risk reduction measures.

asset protection (AS(IEC)61511 term) - function allocated to system design for the purpose of preventing loss to assets.

automatic monitoring (AS4024.1 term) - a back-up safety function which ensures that a safety measure is initiated if the ability of a component or an element to perform its function is diminished, or if the process conditions are changed in such a way that hazards are generated. There are two categories of automatic monitoring 'continuous' automatic monitoring, whereby a safety measure is immediately initiated when a failure occurs, and 'discontinuous' automatic monitoring, whereby a safety measure is initiated during a following machine cycle, if a failure has occurred.

availability (general definition) - the probability that the device, equipment or system is operational at any instant in time.

average frequency of a dangerous failure per hour (PFH) (AS(IEC)61508 term) - average frequency of a dangerous failure of an E/E/PE safety related system to perform the specified safety function over a given period of time.

average probability of dangerous failure on demand (PFD_{avg}) (AS(IEC)61508 term) - mean unavailability (see IEC 60050-191) of an E/E/PE safety-related system to perform the specified safety function when a demand occurs from the EUC or EUC control system.

back-up safety functions (AS4024.1 term) - those functions whose failure does not immediately generate a hazard, however, it reduces the level of safety. This covers notably automatic monitoring of any safety critical function (e.g. monitoring of the correct operation of a position switch belonging to an interlocking device).

basic noise emission standard (B-type standard) (AS4024.1 term) - standard which specifies the procedure for determining the noise emission of machinery and equipment in such a way as to obtain reliable, reproducible results with a specified degree of accuracy.

basic process control system (BPCS) (AS(IEC)61511 term) - system which responds to input signals from the process, its associated equipment, other programmable systems and/or an operator and generates output signals causing the process and its associated equipment to operate in the desired manner but which does not perform any safety instrumented functions with a claimed SIL ≥ 1.

beta factor (β) (AS(IEC)61508 term) - the fraction of failures that have a common cause, and so affect all devices in a redundant set. Beta factor is used in PFD and PFH calculations to account for Common Cause Failure (CCF).

category (CAT) (AS4024.1 term) - classification of the safety-related parts of a control system in respect of their resistance to faults and their subsequent behaviour in the fault condition, and which is achieved by the structural arrangement of parts and/or by their reliability.

category (CAT) (ISO13849 term) - classification of the safety-related parts of a control system in respect of their resistance to faults and their subsequent behaviour in the fault condition, and which is achieved by the structural arrangement of the parts, fault detection and/or by their reliability.

channel (AS(IEC)61508 term) - element or group of elements that independently implement an element safety function.

channel (AS(IEC)61511 term) - element or group of elements that independently perform(s) a function.

coding (AS(IEC)61511 term) - see "programming".

common cause failure (CCF) (AS(IEC)61508 term) - failure, that is the result of one or more events, causing concurrent failures of two or more separate channels in a multiple channel system, leading to system failure.

common cause failure (CCF) (AS(IEC)61511 term) - failure, which is the result of one or more events, causing failures of two or more separate channels in a multiple channel system, leading to system failure.

common cause failure (CCF) (AS(IEC)62061 term) - failure, which is the result of one or more events, causing coincident failures of two or more separate channels in a multiple channel (redundant architecture) subsystem, leading to failure of a SRCF random hardware failure.

common cause failure (CCF) (ISO13849 term) - failures of different items, resulting from a single event, where these failures are not consequences of each other.

common cause failures (AS4024.1 term) - failures of different items, resulting from a single event, where these failures are not consequences of each other.

common mode failure (AS(IEC)61511 term) - failure of two or more channels in the same way, causing the same erroneous result.

common mode failures (AS4024.1 term) - failures of items characterized by the same fault mode.

complex component (AS(IEC)62061 term) - component in which the failure modes are not well-defined or the behaviour under fault conditions cannot be completely defined.

component (AS(IEC)61511 term) - one of the parts of a system, subsystem, or device performing a specific function.

configuration (AS(IEC)61511 term) - see "architecture".

configuration baseline (AS(IEC)61508 term) - information that allows the software release to be recreated in an auditable and systematic way, including: all source code, data, run time files, documentation, configuration files, and installation scripts that comprise a software release; information about compilers, operating systems, and development tools used to create the software release.

configuration management (AS(IEC)61508 term) - discipline of identifying the components of an evolving system for the purposes of controlling changes to those components and maintaining continuity and traceability throughout the lifecycle.

configuration management (AS(IEC)61511 term) - discipline of identifying the components of an evolving (hardware and software) system for the purposes of controlling changes to those components and maintaining continuity and traceability throughout the life cycle.

continuous mode safety instrumented function (AS(IEC)61511 term) - where in the event of a dangerous failure of the safety instrumented function a potential hazard will occur without further failure unless action is taken to prevent it.

control circuit (of a machine) (AS4024.1 term) - a circuit used for the operational control of the machine and for protection of the power circuits.

control device (AS4024.1 term) - a device connected into the control circuit and used for controlling the operation of the machine (e.g. position sensor, manual control switch, relay, magnetically operated valve).

control function (AS(IEC)62061 term) - function that evaluates input information or signals and produces output information or activities.

control guard (AS4024.1 term) - guard associated with an interlocking device (with or without guard locking) so that (a) the hazardous machine functions 'covered' by the guard cannot operate until the guard is closed; and (b) closing the guard initiates operation of the hazardous machine function(s).

control system (AS(IEC)61511 term) - system which responds to input signals from the process and/or from an operator and generates output signals causing the process to operate in the desired manner.

controlled stop (AS4024.1 term) - the stopping of machine motion, by for example, reducing the electrical command signal to zero once the stop signal has been recognized by the control but retaining electrical power to the machine actuators during the stopping process.

crushing zone (AS4024.1 term) - zone in which the human body or parts of the human body are exposed to a crushing hazard. This hazard will be generated when (a) two movable parts are moving towards one another; or (b) one movable part is moving toward a fixed part.

danger zone (AS4024.1 term) - any zone within and/or around machinery in which a person is exposed to risk of injury or damage to health.

dangerous failure (AS(IEC)61511 term) - failure which has the potential to put the safety instrumented system in a hazardous or fail-to-function state.

dangerous failure (AS(IEC)61508 term) - failure of an element and/or subsystem and/or system that plays a part in implementing the safety function that: a) prevents a safety function from operating when required (demand mode) or causes a safety function to fail (continuous mode) such that the EUC is put into a hazardous or potentially hazardous state; or b) decreases the probability that the safety function operates correctly when required.

dangerous failure (AS(IEC)62061 term) - failure of a SRECS, a subsystem, or a subsystem element that has the potential to cause a hazard or non-functional state.

dangerous failure (ISO13849 term) - failure which has the potential to put the SRP/CS in a hazardous or fail-to-function state.

data (AS(IEC)61508 term) - information represented in a manner suitable for communication, interpretation, or processing by computers.

degraded state (AS4024.1 term) - the state of an item where that item continues to perform a function at limits lower than the rated values or continues to perform only part of its functions.

demand (AS(IEC)62061 term) - event that causes the SRECS to perform its SRCF.

demand mode safety instrumented function (AS(IEC)61511 term) - where a specified action (for example, closing of a valve) is taken in response to process conditions or other demands. In the event of a dangerous failure of the safety instrumented function a potential hazard only occurs in the event of a failure in the process or the BPCS.

demand rate (r_d) (ISO13849 term) - frequency of demands for a safety-related action of the SRP/CS.

dependent failure (AS(IEC)61508 term) - failure whose probability cannot be expressed as the simple product of the unconditional probabilities of the individual events that caused it.

dependent failure (AS(IEC)61511 term) - failure whose probability cannot be expressed as the simple product of the unconditional probabilities of the individual events which caused it.

design of a machine (AS4024.1 term) - a series of actions including: (a) The study of the machine itself, taking into account all phases of its 'life', i.e. (i) construction; (ii) transport and commissioning; (iii) assembly, installation; and (iv) adjustment. (b) Use, including (i) setting, teaching/programming or process changeover; (ii) operation; (iii) cleaning; (iv) fault finding; and (v) maintenance. (c) De-commissioning, dismantling and, as far as safety is concerned, disposal. (d) The drafting of the instructions relating to all above-mentioned phases of the 'life' of the machine (except construction), dealt with in AS 4024.1201.

detected / revealed / overt (AS(IEC)61508 term) - in relation to hardware, detected by the diagnostic tests, proof tests, operator intervention (for example physical inspection and manual tests), or through normal operation.

detected / revealed / overt (AS(IEC)61511 term) - in relation to hardware failures and software faults, detected by the diagnostic tests or through normal operation.

deterring device / impeding device (AS4024.1 term) - any physical obstacle which, without totally preventing access to a danger zone, reduces the probability of access to this zone by offering an obstruction to free access.

device (AS(IEC)61511 term) - functional unit of hardware or software, or both, capable of accomplishing a specified purpose (for example, field devices; equipment connected to the field side of the SIS I/O terminals; such equipment includes field wiring, sensors, final elements, logic solvers, and those operator interface devices hard-wired to SIS I/O terminals).

diagnostic coverage (DC) (AS(IEC)62061 term) - decrease in the probability of dangerous hardware failures resulting from the operation of the automatic diagnostic tests.

diagnostic coverage (DC) (AS(IEC)61508 term) - fraction of dangerous failures detected by automatic on-line diagnostic tests. The fraction of dangerous failures is computed by using the dangerous failure rates associated with the detected dangerous failures divided by the total rate of dangerous failures *(DC can be calculated using the following equation:* $DC = \Sigma\lambda_{DD} / \Sigma \lambda_D$*, where* λ_{DD} *is the rate of detected dangerous hardware failures and* λ_D *is the rate of total dangerous hardware failures).*

diagnostic coverage (DC) (AS(IEC)61511 term) - ratio of the detected failure rate to the total failure rate of the component or subsystem as detected by diagnostic tests. Diagnostic coverage does not include any faults detected by proof tests.

diagnostic coverage (DC) (ISO13849 term) - measure of the effectiveness of diagnostics, which may be determined as the ratio between the failure rate of detected dangerous failures and the failure rate of total dangerous failures.

diagnostic test interval (AS(IEC)61508 term) - interval between on-line tests to detect faults in a safety-related system that has a specified diagnostic coverage.

direct contact (AS4024.1 term) - contact of persons or livestock with live parts.

diversity (AS(IEC)61508 term) - different means of performing a required function.

diversity (AS(IEC)61511 term) - existence of different means performing a required function.

dynamic testing (AS(IEC)61508 term) - executing software and/or operating hardware in a controlled and systematic way, so as to demonstrate the presence of the required behaviour and the absence of unwanted behaviour.

E/E/PE system design requirements specification (AS(IEC)61508 term) - specification containing the design requirements for the E/E/PE safety-related system in terms of the subsystems and elements.

E/E/PE system safety functions requirements specification (AS(IEC)61508 term) - specification containing the requirements for the safety functions that have to be performed by the safety-related systems.

E/E/PE system safety integrity requirements specification (AS(IEC)61508 term) - specification containing the safety integrity requirements of the safety functions that have to be performed by the safety-related systems.

E/E/PE system safety requirements specification (AS(IEC)61508 term) - specification containing the requirements for the safety functions and their associated safety integrity levels.

electrical control system (AS(IEC)62061 term) - all the electrical, electronic and programmable electronic parts of the machine control system used to provide, for example, operational control, monitoring, interlocking, communications, protection and safety-related control functions.

electrical/electronic/programmable (E/E/PE) (AS(IEC)61511 term) - based on electrical (E) and/or electronic (E) and/or programmable electronic (PE) technology.

electrical/electronic/programmable electronic (E/E/PE) (AS(IEC)61508 term) - based on electrical (E) and/or electronic (E) and/or programmable electronic (PE) technology.

electrical/electronic/programmable electronic system (E/E/PES) (AS(IEC)61508 term) - system for control, protection or monitoring based on one or more electrical/electronic programmable electronic (E/E/PE) devices, including all elements of the system such as power supplies, sensors and other input devices, data highways and other communication paths, and actuators and other output devices.

element (AS(IEC)61508 term) - part of a subsystem comprising a single component or any group of components that performs one or more element safety functions.

element safety function (AS(IEC)61508 term) - that part of a safety function which is implemented by an element.

embedded software (AS(IEC)61511 term) - software that is part of the system supplied by the manufacturer and is not accessible for modification by the end-user. Embedded software is also referred to as firmware or system software.

embedded software (AS(IEC)62061 term) - software, supplied by the manufacturer, that is part of the SRECS and that is not normally accessible for modification.

embedded software (ISO13849 term) – firmware, system software, software that is part of the system supplied by the control manufacturer and which is not accessible for modification by the user of the machinery.

emergency stop (function) (AS4024.1 term) - function which is intended (a) to avert arising or to reduce existing hazards to persons, damage to machinery or to work in progress; and (b) to be initiated by a single human action when the normal stopping function is inadequate for this purposes. Hazards for the purpose of this standard are those which may arise from (i) functional irregularities (e.g. malfunctioning of the machinery, unacceptable properties of the processed material, human errors); or (ii) normal operation.

emergency stop equipment (AS4024.1 term) - arrangement of components intended to achieve the emergency stop function.

emission sound pressure (p) (AS4024.1 term) - the sound pressure, at a specified position near a noise source, when the source is in operation under specified operating and mounting conditions on a reflecting plane surface, excluding the effects of background noise and of reflections other than those from the plane or planes permitted for the purposes of the test. It is expressed in pascals.

enabling (control) device (AS4024.1 term) - additional manually operated control device used in conjunction with a start control and which, when continuously actuated, allows a machine to function.

enclosure (AS4024.1 term) - a part providing protection of equipment against certain external influences and, in any direction protection against direct contact. An enclosure may be: (a) a cabinet or box, either mounted on the machine or separate from the machine; (b) a compartment consisting of an enclosed space within the machine structure.

environment (AS(IEC)61508 term) - all relevant parameters that can affect the achievement of functional safety in the specific application under consideration and in any safety lifecycle phase.

equipment under control (EUC) (AS(IEC)61508 term) - equipment, machinery, apparatus or plant used for manufacturing, process, transportation, medical or other activities.

equipotential bonding (AS4024.1 term) - electrical connection putting various exposed conductive parts and extraneous conductive parts at a substantially equal potential.

error (AS(IEC)61508 term) - discrepancy between a computed, observed or measured value or condition and the true, specified or theoretically correct value or condition.

error (AS(IEC)61511 term) - discrepancy between a computed, observed or measured value or condition and the true, specified or theoretically correct value or condition.

EUC control system (AS(IEC)61508 term) - system that responds to input signals from the process and/or from an operator and generates output signals causing the EUC to operate in the desired manner.

EUC risk (AS(IEC)61508 term) - risk arising from the EUC or its interaction with the EUC control system.

exposed conductive part (AS4024.1 term) - a conductive part of electrical equipment, which can be touched and which is not normally live, but which may become live under fault conditions.

external risk reduction facilities (AS(IEC)61511 term) - measures to reduce or mitigate the risks, which are separate and distinct from the SIS.

extraneous conductive part (AS4024.1 term) - a conductive part not forming part of the electrical installation and liable to introduce a potential, generally the earth potential.

failure (AS(IEC)61508 term) - termination of the ability of a functional unit to provide a required function or operation of a functional unit in any way other than as required.

failure (AS(IEC)61511 term) - termination of the ability of a functional unit to perform a required function.

failure (AS(IEC)62061 term) - termination of the ability of a SRECS, a subsystem, or a subsystem element to perform a required function.

failure (AS4024.1 term) - the termination of the ability of an item to perform a required function.

failure (ISO13849 term) - termination of the ability of an item to perform a required function.

failure rate (AS(IEC)61508 term) - reliability parameter ($\lambda(t)$) of an entity (single components or systems) such that $\lambda(t).dt$ is the probability of failure of this entity within $[t, t+dt]$ provided that it has not failed during $[0, t]$.

failure to danger (AS4024.1 term) - any failure in the machinery, or in its power supply, that generates a hazardous situation.

fault (AS(IEC)61508 term) - abnormal condition that may cause a reduction in, or loss of, the capability of a functional unit to perform a required function.

fault (AS(IEC)61511 term) - abnormal condition that may cause a reduction in, or loss of, the capability of a functional unit to perform a required function.

fault (AS(IEC)62061 term) - abnormal condition that may cause a reduction in or loss of, the capability of a SRECS, a subsystem, or a subsystem element to perform a required function.

fault (AS4024.1 term) - the state of an item characterized by inability to perform a required function, excluding the inability during preventative maintenance or other planned actions, or due to lack of external resources.

fault (ISO13849 term) - state of an item characterized by the inability to perform a required function, excluding the inability during preventive maintenance or other planned actions, or due to lack of external resources.

fault avoidance (AS(IEC)61508 term) - use of techniques and procedures that aim to avoid the introduction of faults during any phase of the safety lifecycle of the safety-related system.

fault avoidance (AS(IEC)61511 term) - use of techniques and procedures which aim to avoid the introduction of faults during any phase of the safety life cycle of the safety instrumented system.

fault tolerance (AS(IEC)61508 term) - ability of a functional unit to continue to perform a required function in the presence of faults or errors.

fault tolerance (AS(IEC)61511 term) - ability of a functional unit to continue to perform a required function in the presence of faults or errors.

fault tolerance (AS(IEC)62061 term) - ability of a SRECS, a subsystem, or subsystem element to continue to perform a required function in thc presence of faults or failures.

final element (AS(IEC)61511 term) - part of a safety instrumented system which implements the physical action necessary to achieve a safe state.

fixed guard (AS4024.1 term) - guard kept in place by (i.e. closed) either (a) permanently (by welding, etc); or (b) by means of fasteners (screws, nuts, etc.) making removal/opening impossible without using tools.

fixed program language (FPL) (AS(IEC)61511 term) - in this type of language, the user is limited to adjustment of a few parameters (for example, range of the pressure transmitter, alarm levels, network addresses).

full variability language (FVL) (AS(IEC)61511 term) - this type of language is designed to be comprehensible to computer programmers and provides the capability to implement a wide variety of functions and applications.

full variability language (FVL) (AS(IEC)62061 term) - type of language that provides the capability to implement a wide variety of functions and applications.

full variability language (FVL) (ISO13849 term) - type of language that provides the capability of implementing a wide variety of functions and applications.

function block (AS(IEC)62061 term) - smallest element of a SRCF whose failure can result in a failure of the SRCF.

function block element (AS(IEC)62061 term) - part of a function block.

functional safety (AS(IEC)61508 term) - part of the overall safety relating to the EUC and the EUC control system that depends on the correct functioning of the E/E/PE safety-related systems and other risk reduction measures.

functional safety (AS(IEC)61511 term) - part of the overall safety relating to the process and the BPCS which depends on the correct functioning of the SIS and other protection layers.

functional safety (AS(IEC)62061 term) - part of the safety of the machine and the machine control system which depends on the correct functioning of the SRECS, "other technology" safety-related systems and external risk reduction facilities.

functional safety assessment (AS(IEC)61508 term) - investigation, based on evidence, to judge the functional safety achieved by one or more E/E/PE safety-related systems and/or other risk reduction measures.

functional safety assessment (AS(IEC)61511 term) - investigation, based on evidence, to judge the functional safety achieved by one or more protection layers.

functional safety audit (AS(IEC)61508 term) - systematic and independent examination to determine whether the procedures specific to the functional safety requirements to comply with the planned arrangements are implemented effectively and are suitable to achieve the specified objectives.

functional safety audit (AS(IEC)61511 term) - systematic and independent examination to determine whether the procedures specific to the functional safety requirements comply with the planned arrangements, are implemented effectively and are suitable to achieve the specified objectives.

functional unit (AS(IEC)61508 term) - entity of hardware or software, or both, capable of accomplishing a specified purpose.

functional unit (AS(IEC)61511 term) - entity of hardware or software, or both, capable of accomplishing a specified purpose.

guard (AS4024.1 term) - part of a machine specifically used to provide protection by means of a physical barrier. Depending on its construction, a guard may be called casing, cover, screen, door, enclosing guard, etc.

guard locking device (AS4024.1 term) - device intended to lock a guard in the closed position and linked to the control system so that (a) the machine cannot operated until the guard is closed and locked; and (b) the guard remains locked until the risk has passed.

hardware fault tolerance (HWFT or HFT) (AS(IEC)61508 term) - a numerical measure of the level of fault tolerance in a system. A safety system with HWFT = n, requires n+1 failures in order to be defeated (eg. to defeat a redundant 1-out-of-2 system, two (2) failures are required. Therefore, HWFT =1).

hardware safety integrity (AS(IEC)61508 term) - part of the safety integrity of a safety-related system relating to random hardware failures in a dangerous mode of failure.

hardware safety integrity (AS(IEC)61511 term) - part of the safety integrity of the safety instrumented function relating to random hardware failures in a dangerous mode of failure.

hardware safety integrity (AS(IEC)62061 term) - part of the safety integrity of a SRECS or its subsystems comprising requirements for both the probability of dangerous random hardware failures and architectural constraints.

harm (AS(IEC)61508 term) - physical injury or damage to the health of people or damage to property or the environment.

harm (AS(IEC)61511 term) - physical injury or damage to the health of people, either directly or indirectly, as a result of damage to property or to the environment.

harm (AS4024.1 term) - physical injury and/or damage to health or property.

harm (ISO13849 term) - physical injury or damage to health.

harmful event (AS(IEC)61508 term) - occurrence in which a hazardous situation or hazardous event results in harm.

hazard (AS(IEC)61508 term) - potential source of harm.

hazard (AS(IEC)61511 term) - potential source of harm.

hazard (AS4024.1 term) - a source of possible injury or damage to health.

hazard (from machinery) (AS(IEC)62061 term) - potential source of physical injury or damage to health.

hazard (ISO13849 term) - potential source of harm.

hazardous event (AS(IEC)61508 term) - event that may result in harm.

hazardous event (AS4024.1 term) - event that can cause harm.

hazardous machine function (AS4024.1 term) - any function of a machine which generates a hazard when operating.

hazardous situation (AS(IEC)61508 term) -circumstance in which people, property or the environment are exposed to one or more hazards.

hazardous situation (AS(IEC)62061 term) - circumstance in which a person is exposed to a hazard(s).

hazardous situation (AS4024.1 term) - any situation in which a person is exposed to a hazard or to hazards.

hazardous situation (ISO13849 term) - circumstance in which a person is exposed to at least one hazard, the exposure having immediately or over a long period of time the potential to result in harm.

hazardous substances (AS4024.1 term) - any chemical or biological agent which is hazardous to health, e.g. substances or preparations classified as (a) very toxic; (b) toxic; (c) harmful; (d) corrosive; (e) irritant; (f) sensitizing; (g) carcinogenic; (h) mutagenic; (i) teratogenic; (j) pathogenic; or (k) asphyxiants.

high demand or continuous mode (AS(IEC)62061 term) - mode of operation in which the frequency of demands on a SRECS is greater than one per year or greater than twice the proof-test frequency.

hold-to-run control device (AS4024.1 term) - control device which initiates and maintains operation of machine elements only as long as the manual control (actuator) is actuated. The manual control (actuator) returns automatically to the stop position when released.

human error (AS(IEC)61511 term) - mistake, human action or inaction that produces an unintended result.

impact analysis (AS(IEC)61508 term) - activity of determining the effect that a change to a function or component in a system will have to other functions or components in that system as well as to other systems.

impact analysis (AS(IEC)61511 term) - activity of determining the effect that a change to a function or component will have to other functions or components in that system as well as to other systems.

independent department (AS(IEC)61508 term) - department that is separate and distinct from the departments responsible for the activities which take place during the specific phase of the overall, E/E/PE system or software safety lifecycle that is subject to the functional safety assessment or validation.

independent department (AS(IEC)61511 term) - department which is separate and distinct from the departments responsible for the activities which take place during the specific phase of the safety life cycle that is subject to the functional safety assessment or validation.

independent organisation (AS(IEC)61508 term) - organisation that is separate and distinct, by management and other resources, from the organisations responsible for the activities that take place during the specific phase of the overall, E/E/PE system or software safety lifecycle that is subject to the functional safety assessment or validation.

independent organization (AS(IEC)61511 term) - organization which is separate and distinct, by management and other resources, from the organizations responsible for the activities which take place during the specific phase of the safety life cycle that is subject to the functional safety assessment or validation.

independent person (AS(IEC)61508 term) - person who is separate and distinct from the activities which take place during the specific phase of the overall, E/E/PE system or software safety lifecycle that is subject to the functional safety assessment or validation, and does not have direct responsibility for those activities.

independent person (AS(IEC)61511 term) - person who is separate and distinct from the activities which take place during the specific phase of the safety life cycle that is subject to the functional safety assessment or validation and does not have direct responsibility for those activities.

indirect contact (AS4024.1 term) - contact of persons or livestock with exposed conductive parts which have become live under fault conditions.

information for use (AS4024.1 term) - safety measures consisting of communication inks, such as texts, words, signs, signals, symbols or diagrams, used separately or in combination, to convey information to the user. It is directed to professional and/or non-professional users.

input function (AS(IEC)61511 term) - function which monitors the process and its associated equipment in order to provide input information for the logic solver.

instrument (AS(IEC)61511 term) - apparatus used in performing an action (typically found in instrumented systems).

intended use of a machine (AS4024.1 term) - the use for which the machine is suited according to the information provided by the manufacturer or which is deemed usual according to its design, construction and function. Intended use also involves the compliance with the technical instructions laid down notably in the instruction handbook taking into account of reasonably foreseeable misuse.

intended use of a machine (ISO13849 term) - use of the machine in accordance with the information provided in the instructions for use.

interlocking devices / interlock (AS4024.1 term) - mechanical, electrical or other types of device, the purpose of which is to prevent the operation of machine elements under specified conditions, (generally as long as a guard is not closed).

interlocking guard (AS4024.1 term) - guard associated with an interlocking device, so that (a) the hazardous machine functions 'covered' by the guard cannot operate until the guard is closed; (b) if the guard is opened while the hazardous machine functions are operating, a stop instruction is given; and (c) when the guard is closed, the hazardous machine functions 'covered' by the guard can operate, but the closure of the guard does not by itself initiate their operation.

interlocking guard with guard locking (AS4024.1 term) - guard associated with an interlocking device and a guard locking device so that (a) the hazardous machine functions 'covered' by the guard cannot operate until the guard is closed and locked; (b) the guard remains closed and locked until the risk of injury from the hazardous machine function has passed; and (c) when the guard is closed and locked, the hazardous machine functions 'covered' by the guard can operate but the closure and locking of the guard do not by themselves initiate their operation.

isolation and energy dissipation (AS4024.1 term) - a procedure which consists of all of the four following actions: (a) isolating (disconnecting, separating) the machine (or defined parts of the machine) from all power supplies. (b) if necessary (for instance in large machines or in installations), locking (or otherwise securing) all the isolation units in the isolating position. (c) dissipating or restraining (containing) any stored energy which may give rise to a hazard.

limited movement control device (AS4024.1 term) - control device, the actuation of which permits only a limited amount of travel of a machine element thus minimizing the risk as much as possible; further movement is precluded until there is a subsequent and separate actuation of the control.

limited variability language (AS(IEC)61508 term) - software programming language, whose notation is textual or graphical or has characteristics of both, for commercial and industrial programmable electronic controllers with a range of capabilities limited to their application.

limited variability language (LVL) (AS(IEC)62061 term) - type of language that provides the capability to combine predefined, application specific, library functions to implement the safety requirements specifications.

limited variability language (LVL) (ISO13849 term) - type of language that provides the capability of combining predefined, application-specific library functions to implement the safety requirements specifications.

limited variability language (LVL)(AS(IEC)61511 term) - this type of language is designed to be comprehensible to process sector users, and provides the capability to combine predefined, application specific, library functions to implement the safety requirements specifications. An LVL provides a close functional correspondence with the functions required to achieve the application.

limiting device (AS4024.1 term) - device which prevents a machine or machine elements from exceeding a designed limit (e.g. space limit, pressure limit, etc).

live part (AS4024.1 term) - a conductor or conductive part intended to be energized in normal use, including a neutral conductor, but, by convention, not a PEN conductor.

logic function (AS(IEC)61511 term) - function which performs the transformations between input information (provided by one or more input functions) and output information (used by one or more output functions); logic functions provide the transformation from one or more input functions to one or more output functions.

logic solver (AS(IEC)61511 term) - that portion of either a BPCS or SIS that performs one or more logic function(s).

low complexity component (AS(IEC)62061 term) - component in which the failure modes are well-defined and the behaviour under fault conditions can be completely defined.

low complexity E/E/PE safety-related system (AS(IEC)61508 term) - E/E/PE safety-related system in which the failure modes of each individual component are well defined and the behaviour of the system under fault conditions can be completely determined.

low demand mode (AS(IEC)62061 term) - mode of operation in which the frequency of demands on a SRECS is no greater than one per year and no greater than twice the proof-test frequency.

machine (machinery) (AS4024.1 term) - an assembly of linked parts or components, at least one of which moves, with the appropriate machine actuators, control and power circuits, etc., joined together for a specific application, in particular for the processing, treatment, moving or packaging of a material. The term machinery also covers an assembly of machines which, in order to achieve one and the same end, are arranged and controlled so that they function as an integral whole.

machine actuator (AS4024.1 term) - a power mechanism used to effect motion of the machine.

machine control system (AS(IEC)62061 term) - system which responds to an input from, for example, the process, other machine elements, an operator, external control equipment, and generates an output(s) causing the machine to behave in the intended manner.

machine control system (ISO13849 term) - system which responds to input signals from parts of machine elements, operators, external control equipment or any combination of these and generates output signals causing the machine to behave in the intended manner.

machinery (AS(IEC)62061 term) - assembly of linked parts or components, at least one of which moves, with the appropriate machine actuators, control and power circuits, joined together for a specific application, in particular for the processing, treatment, moving or packaging of a material. The terms “machinery” and “machine” also cover an assembly of machines which, in order to achieve the same end, are arranged and controlled so that they function as an integral whole.

maintainability of a machine (AS4024.1 term) - the ability of a machine to be maintained in a state which enables it to fulfil its function under conditions of intended use or restored into such a state, the necessary actions (maintenance) being carried out according to specified practices and using specified means.

maintenance/engineering interface (AS(IEC)61511 term) - maintenance/engineering interface is that hardware and software provided to allow proper SIS maintenance or modification. It can include instructions and diagnostics which may be found in software, programming terminals with appropriate communication protocols, diagnostic tools, indicators, bypass devices, test devices, and calibration devices.

manual control actuator (AS4024.1 term) - that component of the control device which, when operated, activates the control device, and is designed to be operated by a person.

manual reset (AS4024.1 term) - function within the safety-related part of the control system to manually restore given safety functions before the re-instating of a machine.

manual reset (ISO13849 term) - function within the SRP/CS used to restore manually one or more safety functions before re-starting a machine.

mean repair time (MRT) (AS(IEC)61508 term) - expected overall repair time.

mean time to dangerous failure ($MTTF_d$) (ISO13849 term) - expectation of the mean time to dangerous failure.

mean time to failure (MTTF) (AS(IEC)62061 term) - expectation of the mean time to failure.

mean time to restoration (MTTR) (AS(IEC)61508 term) - expected time to achieve restoration.

mechanical restraint device (AS4024.1 term) - device which introduces into a mechanism, a mechanical obstacle, (wedge, spindle, strut, scotch, etc.) which, by virtue of its own strength, can prevent any hazardous movement (for instance, the fall of a ram due to the failure of the normal retaining system).

mission time (T_M) (ISO13849 term) - period of time covering the intended use of an SRP/CS.

mitigation (AS(IEC)61511 term) - action that reduces the consequence(s) of a hazardous event.

mode of operation (AS(IEC)61508 term) - way in which a safety function operates, which may be either: low demand mode - where the safety function is only performed on demand, in order to transfer the EUC into a specified safe state, and where the frequency of demands is no greater than one per year; high demand mode - where the safety function is only performed on demand, in order to transfer the EUC into a specified safe state, and where the frequency of demands is greater than one per year; or continuous mode - where the safety function retains the EUC in a safe state as part of normal operation.

mode of operation (AS(IEC)61511 term) - way in which a safety instrumented function operates.

module (AS(IEC)61511 term) - self-contained assembly of hardware components that performs a specific hardware function (i.e. digital input module, analogue output module), or reusable application program (can be internal to a program or a set of programs) that support a specific function, for example, portion of a computer program that carries out a specific function.

monitoring (ISO13849 term) - safety function which ensures that a protective measure is initiated if the ability of a component or an element to perform its function is diminished or if the process conditions are changed in such a way that a decrease of the amount of risk reduction is generated.

MooN (AS(IEC)61511 term) - safety instrumented system, or part thereof, made up of "N" independent channels, which are so connected, that "M" channels are sufficient to perform the safety instrumented function.

MTBF (general definition) - Mean Time Between Failures. MTBF = MTTF + MTTR. See 'mean time to failure' and 'mean time to restoration'.

muting (AS4024.1 term) - temporary automatic suspension of a safety function(s) by safety-related parts of the control system.

muting (ISO13849 term) - temporary automatic suspension of a safety function(s) by the SRP/CS.

necessary risk reduction (AS(IEC)61508 term) - risk reduction to be achieved by the E/E/PE safety-related systems and/or other risk reduction measures in order to ensure that the tolerable risk is not exceeded.

necessary risk reduction (AS(IEC)61511 term) - risk reduction required to ensure that the risk is reduced to a tolerable level.

neutral conductor (symbol N) (AS4024.1 term) - a conductor connected to the neutral point of a system and capable of contributing to the transmission of electrical energy.

no effect failure (AS(IEC)61508 term) - failure of an element that plays a part in implementing the safety function but has no direct effect on the safety function.

no part failure (AS(IEC)61508 term) - failure of a component that plays no part in implementing the safety function.

noise emission declaration (AS4024.1 term) - information on the noise emitted by the machine, given by the manufacturer or supplier in technical documents or other literature concerning noise emission values. The noise emission declaration may take the form of either the declared single-number noise emission value or the declared dual-number noise emission value.

noise test code (C-type standard) (AS4024.1 term) - a standard that is applicable to a particular class, family or type of machinery or equipment, which specifies all the information necessary to carry out efficiently the determination, declaration and verification of the noise emission characteristics under standardized conditions.

non-programmable (NP) system (AS(IEC)61511 term) - system based on non-computer technologies (i.e. a system not based on programmable electronics (PE) or software).

operator (AS4024.1 term) - the person or persons given the task of installing, operating, adjusting, maintaining, cleaning, repairing, or transporting machinery.

operator interface (AS(IEC)61511 term) - means by which information is communicated between a human operator(s) and the SIS (for example, CRTs, indicating lights, push-buttons, horns, alarms); the operator interface is sometimes referred to as the human-machine interface (HMI).

other risk reduction measure (AS(IEC)61508 term) - measure to reduce or mitigate risk that is separate and distinct from, and does not use, E/E/PE safety-related systems.

other technology safety related systems (AS(IEC)61511 term) - safety related systems that are based on a technology other than electrical, electronic, or programmable electronic.

output function (AS(IEC)61511 term) - function which controls the process and its associated equipment according to final actuator information from the logic function.

overall safety function (AS(IEC)61508 term) - means of achieving or maintaining a safe state for the EUC, in respect of a specific hazardous event.

overcurrent (AS4024.1 term) - any current exceeding the rated value. For conductors, the rated value is the current carrying capacity.

overload (of a circuit) (AS4024.1 term) - the time/current relationship in a circuit which is in excess of the rated full load of the circuit when the circuit is not under a fault condition.

performance level (PL) (ISO13849 term) - discrete level used to specify the ability of safety-related parts of control systems to perform a safety function under foreseeable conditions.

phase (AS(IEC)61511 term) - period within the safety life cycle where activities described in this standard take place.

PLC (general definition): Programmable Logic Controller.

positive mode actuation (AS4024.1 term) - it a moving mechanical component inevitability moves another component along with it, either by direct contact or via rigid elements, the second component is said to be actuated in the positive mode (or positively) by the first one.

positive opening operation of a contact element (AS4024.1 term) - the achievement of a contact separation as the direct result of a specified movement of the switch actuator through non-resilient members (e.g. not dependent upon springs).

power circuit (AS4024.1 term) - a circuit used for supply power from the supply network to units of equipment used for productive operation and to transformers supplying control circuits.

pre-existing software (AS(IEC)61508 term) - software element which already exists and is not developed specifically for the current project or safety-related system.

prevention (AS(IEC)61511 term) - action that reduces the frequency of occurrence of a hazardous event.

prior use (AS(IEC)61511 term) - see "proven-in-use".

probability of dangerous failure on demand (PFD) (AS(IEC)61508 term) - safety unavailability (see IEC 60050-191) of an E/E/PE safety-related system to perform the specified safety function when a demand occurs from the EUC or EUC control system.

probability of dangerous failure per hour (PFHD) (AS(IEC)62061 term) - average probability of dangerous failure within one (1) hour.

process risk (AS(IEC)61511 term) - risk arising from the process conditions caused by abnormal events (including BPCS malfunction).

process safety time (AS(IEC)61508 term) - period of time between a failure, that has the potential to give rise to a hazardous event, occurring in the EUC or EUC control system and the time by which action has to be completed in the EUC to prevent the hazardous event occurring.

programmable electronic (PE) (AS(IEC)61508 term) - based on computer technology which may be comprised of hardware, software, and of input and/or output units.

programmable electronic system (PES) (AS(IEC)61508 term) - system for control, protection or monitoring based on one or more programmable electronic devices, including all elements of the system such as power supplies, sensors and other input devices, data highways and other communication paths, and actuators and other output devices.

programmable electronic system (PES) (ISO13849 term) - system for control, protection or monitoring dependent for its operation on one or more programmable electronic devices, including all elements of the system such as power supplies, sensors and other input devices, contactors and other output devices.

programmable electronics (PE) (AS(IEC)61511 term) - electronic component or device forming part of a PES and based on computer technology. The term encompasses both hardware and software and input and output units.

programming (AS(IEC)61511 term) - process of designing, writing and testing a set of instructions for solving a problem or processing data.

proof test (AS(IEC)61508 term) - periodic test performed to detect dangerous hidden failures in a safety-related system so that, if necessary, a repair can restore the system to an "as new" condition or as close as practical to this condition.

proof test (AS(IEC)61511 term) - test performed to reveal undetected faults in a safety instrumented system so that, if necessary, the system can be restored to its designed functionality.

proof test (AS(IEC)62061 term) - test that can detect faults and degradation in a SRECS and its subsystems so that, if necessary, the SRECS and its subsystems can be restored to an "as new" condition or as close as practical to this condition.

protection layer (AS(IEC)61511 term) - any independent mechanism that reduces risk by control, prevention or mitigation.

protective bonding circuit (AS4024.1 term) - the whole of the protective conductors and conductor parts used for protection against electric shock in the event of an insulation failure.

protective conductor (AS4024.1 term) - a conductor required by some measures for protection against electric shock for electrically connecting any of the following parts: (a) exposed conductive parts. (b) extraneous conductive parts. (c) main earthing terminal.

protective measure (AS(IEC)62061 term) - measure intended to achieve risk reduction

protective measure (ISO13849 term) - measure intended to achieve risk reduction.

protective structure (AS4024.1 term) - a physical obstruction, e.g. guard, part of a machine, which restricts the movement of the body and/or part of it.

proven in use (AS(IEC)61508 term) - demonstration, based on an analysis of operational experience for a specific configuration of an element, that the likelihood of dangerous systematic faults is low enough so that every safety function that uses the element achieves its required safety integrity level.

proven-in-use (AS(IEC)61511 term) - when a documented assessment has shown that there is appropriate evidence, based on the previous use of the component, that the component is suitable for use in a safety instrumented system (see "prior use").

quality (AS(IEC)61511 term) - totality of characteristics of an entity that bear on its ability to satisfy stated and implied needs.

random hardware failure (AS(IEC)61508 term) - failure, occurring at a random time, which results from one or more of the possible degradation mechanisms in the hardware.

random hardware failure (AS(IEC)61511 term) - failure, occurring at a random time, which results from a variety of degradation mechanisms in the hardware.

random hardware failure (AS(IEC)62061 term) - failure occurring at a random time, which results from one or more of the possible degradation mechanisms in the hardware.

reasonably foreseeable misuse (AS(IEC)61508 term) - use of a product, process or service in a way not intended by the supplier, but which may result from readily predictable human behaviour.

reasonably foreseeable misuse (ISO13849 term) - use of a machine in a way not intended by the designer, but which may result from readily predictable human behaviour.

redundancy (AS(IEC)61508 term) - the existence of more than one means for performing a required function or for representing information.

redundancy (AS(IEC)61511 term) - use of multiple elements or systems to perform the same function; redundancy can be implemented by identical elements (identical redundancy) or by diverse elements (diverse redundancy).

redundancy (AS4024.1 term) - the application of more than one device or system, or part of a device or system, with the objective of ensuring that in the event of one failing to perform its function another is available to perform that function.

reliability (general definition): the probability that a device, equipment or system performs a required function under specified conditions for a given period of time.

reliability of a machine (AS4024.1 term) - the ability of a machine or components, or equipment, to perform a required function under specified conditions and for a given period of time without failing.

repair rate (r_r) (ISO13849 term) - reciprocal value of the period of time between detection of a dangerous failure by either an online test or obvious malfunction of the system and the restart of operation after repair or system/component replacement.

required performance level (PL_r) (ISO13849 term) - performance level (PL) applied in order to achieve the required risk reduction for each safety function.

residual risk (AS(IEC)61508 term) - risk remaining after protective measures have been taken.

residual risk (AS4024.1 term) - risk remaining after safety measures have been taken.

residual risk (ISO13849 term) - risk remaining after protective measures have been taken.

risk (AS(IEC)61508 term) - combination of the probability of occurrence of harm and the severity of that harm.

risk (AS(IEC)61511 term) - combination of the frequency of occurrence of harm and the severity of that harm.

risk (AS(IEC)62061 term) - combination of the probability of occurrence of harm and the severity of that harm.

risk (AS4024.1 term) - a combination of the probability and the degree of the possible injury or damage to health in a hazardous situation.

risk (ISO13849 term) - combination of the probability of occurrence of harm and the severity of that harm

risk analysis (ISO13849 term) - combination of the specification of the limits of the machine, hazard identification and risk estimation.

risk assessment (AS4024.1 term) - a comprehensive estimation of the probability and the degree of the possible injury or damage to health in a hazardous situation in order to select appropriate safety measures.

risk assessment (ISO13849 term) - overall process comprising risk analysis and risk evaluation.

risk evaluation (ISO13849 term) - judgement, on the basis of risk analysis, of whether risk reduction objectives have been achieved.

risk reduction by design (AS4024.1 term) - safety measures consisting of (a) avoiding or reducing as many of the hazards as possible by suitable choice of design features; and (b) limiting exposure to hazards which are unavoidable or cannot be reduced sufficiently; this is achieved by reducing the need for operator intervention in danger zones.

safe failure (AS(IEC)61508 term) - failure of an element and/or subsystem and/or system that plays a part in implementing the safety function that: a) results in the spurious operation of the safety function to put the EUC (or part thereof) into a safe state or maintain a safe state; or b) increases the probability of the spurious operation of the safety function to put the EUC (or part thereof) into a safe state or maintain a safe state).

safe failure (AS(IEC)61511 term) - failure which does not have the potential to put the safety instrumented system in a hazardous or fail-to-function state.

safe failure (AS(IEC)62061 term) - failure of a SRECS, a subsystem of a SRECS, or a subsystem element of a SRECS that does not have the potential to cause a hazard.

safe failure fraction (SFF) (AS(IEC)61508 term) - property of a safety related element that is defined by the ratio of the average failure rates of safe plus dangerous detected failures and safe plus dangerous failures *(SFF can be calculated using the following equation:* $SFF = (\Sigma\lambda_S + \Sigma\lambda_{DD}) / (\Sigma\lambda_S + \Sigma\lambda_D)$ *where* λ_S *is the rate of safe failure, and* λ_D *is the rate of dangerous failure and* λ_{DD} *is the rate of dangerous failure which is detected by the diagnostic functions).*

safe failure fraction (SFF) (AS(IEC)61511 term) - fraction of the overall random hardware failure rate of a device that results in either a safe failure or a detected dangerous failure.

safe failure fraction (SFF) (AS(IEC)62061 term) - fraction of the overall failure rate of a subsystem that does not result in a dangerous failure

safe state (AS(IEC)61508 term) - state of the EUC when safety is achieved.

safe state (AS(IEC)61511 term) - state of the process when safety is achieved.

safeguarding (AS4024.1 term) - safety measure consisting of the use of specific technical means called safeguards (guards, safety devices), to protect persons from the hazards which cannot be reasonable be removed or sufficiently limited by design.

safety (AS(IEC)61508 term) - freedom from unacceptable risk.

safety (AS(IEC)61511 term) - freedom from unacceptable risk.

safety configured logic solver (AS(IEC)61511 term) - general purpose industrial grade PE logic solver which is specifically configured for use in safety applications in accordance with AS(IEC)61511.1 Clause 11.5

safety critical functions (AS4024.1 term) - those functions of a machine, the malfunction of which would immediately increase the risk of injury or damage to health. There are two categories of safety critical functions: (a) Safety-specific functions, which are safety critical functions specifically intended to achieve safety. For example (i) function preventing unintended/unexpected start-up (interlocking device associated with a guard,) (ii) single-cycle function; or (iii) two-hand control function. (b) Safety-related functions, which are safety critical functions other than safety-specific functions. For example (i) manual control of a hazardous mechanism during setting phases, with by-passed (muted) safety devices; and (ii) speed or temperature control keeping the machine within safe operating limits.

safety device (AS4024.1 term) - device (other than a guard) which eliminates or reduces risk, alone or associated with a guard.

safety distance (AS4024.1 term) - the minimum distance a protective structure shall be placed from a danger zone.

safety function (AS(IEC)61508 term) - function to be implemented by an E/E/PE safety-related system or other risk reduction measures, that is intended to achieve or maintain a safe state for the EUC, in respect of a specific hazardous event.

safety function (AS(IEC)61511 term) - function to be implemented by an SIS, other technology safety related system or external risk, reduction facilities, which is intended to achieve or maintain a safe state for the process, with respect to a specific hazardous event.

safety function (AS(IEC)62061 term) - function of a machine whose failure can result in an immediate increase of the risk(s).

safety function (ISO13849 term) - function of the machine whose failure can result in an immediate increase of the risk(s).

safety function of control systems (AS4024.1 term) - function initiated by an input signal and processed by the safety-related parts of the control system to enable the machine (as a system) to achieve a safe state.

safety instrumented control function (AS(IEC)61511 term) - safety instrumented function with a specified SIL operating in continuous mode which is necessary to prevent a hazardous condition from arising and/or to mitigate its consequences.

safety instrumented control system (AS(IEC)61511 term) - instrumented system used to implement one or more safety instrumented control functions.

safety instrumented function (SIF) (AS(IEC)61511 term) - safety function with a specified safety integrity level which is necessary to achieve functional safety and which can be either a safety instrumented protection function or a safety instrumented control function.

safety instrumented system (SIS) (AS(IEC)61511 term) - instrumented system used to implement one or more safety instrumented functions. An SIS is composed of any combination of sensor (s), logic solver (s), and final elements(s).

safety integrity (AS(IEC)61508 term) - probability of an E/E/PE safety-related system satisfactorily performing the specified safety functions under all the stated conditions within a stated period of time.

safety integrity (AS(IEC)61511 term) - average probability of a safety instrumented system satisfactorily performing the required safety instrumented functions under all the stated conditions within a stated period of time.

safety integrity (AS(IEC)62061 term) - probability of a SRECS or its subsystem satisfactorily performing the required safety-related control functions under all stated conditions.

safety integrity level (SIL) (AS(IEC)61508 term) - discrete level (one out of a possible four), corresponding to a range of safety integrity values, where safety integrity level 4 has the highest level of safety integrity and safety integrity level 1 has the lowest.

safety integrity level (SIL) (AS(IEC)61511 term) - discrete level (one out of four) for specifying the safety integrity requirements of the safety instrumented functions to be allocated to the safety instrumented systems. Safety integrity level 4 has the highest level of safety integrity; safety integrity level 1 has the lowest.

safety integrity level (SIL) (AS(IEC)62061 term) - discrete level (one out of a possible three) for specifying the safety integrity requirements of the safety-related control functions to be allocated to the SRECS, where safety integrity level three has the highest level of safety integrity and safety integrity level one has the lowest.

safety integrity level (SIL) (ISO13849 term) - discrete level (one out of a possible four) for specifying the safety integrity requirements of the safety functions to be allocated to the E/E/PE safety-related systems, where safety integrity level 4 has the highest level of safety integrity and safety integrity level 1 has the lowest.

safety integrity requirements specification (AS(IEC)61511 term) - specification that contains the safety integrity requirements of the safety instrumented functions that have to be performed by the safety instrumented system(s).

safety life cycle (AS(IEC)61511 term) - necessary activities involved in the implementation of safety instrumented function(s) occurring during a period of time that starts at the concept phase of a project and finishes when all of the safety instrumented functions are no longer available for use.

safety lifecycle (AS(IEC)61508 term) - necessary activities involved in the implementation of safety-related systems, occurring during a period of time that starts at the concept phase of a project and finishes when all of the E/E/PE safety-related systems and other risk reduction measures are no longer available for use.

safety manual (AS(IEC)61511 term) - manual which defines how the device, subsystem or system can be safely applied.

safety manual for compliant items (AS(IEC)61508 term) - document that provides all the information relating to the functional safety of an element, in respect of specified element safety functions, that is required to ensure that the system meets the requirements of IEC 61508 series.

safety measure (AS4024.1 term) - means that eliminates a hazard or reduces a risk.

safety of a machine (AS4024.1 term) - the ability of a machine to perform its function, to be transported, installed, adjusted, maintained, dismantled and disposed of under conditions of intended use, specified in the Instruction handbook (and, in some cases, within a given period of time indicated in the instruction handbook without causing injury or damage to health.

safety of control systems (AS4024.1 term) - ability of safety-related parts of a control system to perform their safety function(s) for a given time according to their specified category.

safety requirements specification (AS(IEC)61511 term) - specification that contains all the requirements of the safety instrumented functions that have to be performed by the safety instrumented systems.

safety software (AS(IEC)61511 term) - software in a safety instrumented system with application, embedded or utility software functionality.

safety-related control function (SRCF) (AS(IEC)62061 term) - control function implemented by a SRECS with a specified integrity level that is intended to maintain the safe condition of the machine or prevent an immediate increase of the risk(s).

safety-related electrical control system (SRECS) (AS(IEC)62061 term) - electrical control system of a machine whose failure can result in an immediate increase of the risk(s).

safety–related part of a control system (SRP/CS) (ISO13849 term) - part of a control system that responds to safety-related input signals and generates safety-related output signals.

safety-related software (AS(IEC)61508 term) - software that is used to implement safety functions in a safety-related system.

safety-related software (AS(IEC)62061 term) - software that is used to implement safety-related control functions in a safety-related system.

safety-related system (AS(IEC)61508 term) - designated system that both implements the required safety functions necessary to achieve or maintain a safe state for the EUC; and is intended to achieve, on its own or with other E/E/PE safety-related systems and other risk reduction measures, the necessary safety integrity for the required safety functions.

sensor (AS(IEC)61511 term) - device or combination of devices, which measure the process condition (for example, transmitters, transducers, process switches, position switches).

SIL claim limit (for a subsystem) (SILCL) (AS(IEC)62061 term) - maximum SIL that can be claimed for a SRECS subsystem in relation to architectural constraints and systematic safety integrity.

soft-error (AS(IEC)61508 term) - erroneous changes to data content but no changes to the physical circuit itself.

software (AS(IEC)61508 term) - intellectual creation comprising the programs, procedures, data, rules and any associated documentation pertaining to the operation of a data processing system.

software (AS(IEC)61511 term) - intellectual creation comprising the programs, procedures, data, rules and any associated documentation pertaining to the operation of a data processing system.

software life cycle (AS(IEC)61511 term) - activities occurring during a period of time that starts when software is conceived and ends when the software is permanently disused.

software lifecycle (AS(IEC)61508 term) - activities occurring during a period of time that starts when software is conceived and ends when the software is permanently decommissioned.

software module (AS(IEC)61508 term) - construct that consists of procedures and/or data declarations and that can also interact with other such constructs.

software off-line support tool (AS(IEC)61508 term) - software tool that supports a phase of the software development lifecycle and that cannot directly influence the safety-related system during its run time. Software off-line tools may be divided into the following classes: T1 - generates no outputs which can directly or indirectly contribute to the executable code (including data) of the safety related system; T2 - supports the test or verification of the design or executable code, where errors in the tool can fail to reveal defects but cannot directly create errors in the executable software; T3 - generates outputs which can directly or indirectly contribute to the executable code of the safety related system.

software on-line support tool (AS(IEC)61508 term) - software tool that can directly influence the safety-related system during its run time.

software safety integrity (AS(IEC)61508 term) - part of the safety integrity of a safety-related system relating to systematic failures in a dangerous mode of failure that are attributable to software.

software safety integrity (AS(IEC)62061 term) - part of the systematic safety integrity of a SRECS or its subsystems related to the capability of software in a programmable electronic system performing its safety-related control functions under all stated conditions during a stated period of time.

software safety integrity level (AS(IEC)61508 term) - systematic capability of a software element that forms part of a subsystem of a safety-related system.

software template (AS(IEC)61511 term) - structured non-specific piece of application software that can be easily altered to support specific functions while retaining the original structure; for example, an interactive screen template controls the process flow of the application screens, but is not specific to the data being presented; a programmer may take the generic template and make function-specific revisions to produce a new screen for the users.

sound power (W) (AS4024.1 term) - the rate per unit at which airborne sound energy is radiated by a source. It is expressed in watts.

SRECS diagnostic function (AS(IEC)62061 term) - function intended to detect faults in the SRECS and produce a specified output information or activity when a fault is detected.

SRECS fault reaction function (AS(IEC)62061 term) - function that is initiated when a fault within a SRECS is detected by the SRECS diagnostic function.

start-up (machine start-up) (AS4024.1 term) - change from rest to motion of a machine or of one of its parts.

stopping time (time for hazard elimination) (AS4024.1 term) - the period between the point at which the interlocking device initiates the stop command and the point at which the risk from hazardous machine functions has passed.

subsystem (AS(IEC)61508 term) - entity of the top-level architectural design of a safety-related system where a dangerous failure of the subsystem results in dangerous failure of a safety function.

subsystem (AS(IEC)61511 term) - see "system".

subsystem (AS(IEC)62061 term) (AS(IEC)62061 term) - entity of the top-level architectural design of the SRECS where a failure of any subsystem will result in a failure of a safety-related control function.

subsystem element (AS(IEC)62061 term) - part of a subsystem, comprising a single component or any group of components.

switching device (AS4024.1 term) - a device designed to make or break the current in one or more electric circuits.

system (AS(IEC)61511 term) - set of elements, which interact according to a design; an element of a system can be another system, called a subsystem, which may be a controlling system or a controlled system and may include hardware, software and human interaction.

system software (AS(IEC)61508 term) - part of the software of a PE system that relates to the functioning of, and services provided by, the programmable device itself, as opposed to the application software that specifies the functions that perform a task related to the safety of the EUC.

systematic capability (AS(IEC)61508 term) - measure (expressed on a scale of SC 1 to SC 4) of the confidence that the systematic safety integrity of an element meets the requirements of the specified SIL, in respect of the specified element safety function, when the element is applied in accordance with the instructions specified in the compliant item safety manual for the element.

systematic failure (AS(IEC)61508 term) - failure, related in a deterministic way to a certain cause, which can only be eliminated by a modification of the design or of the manufacturing process, operational procedures, documentation or other relevant factors.

systematic failure (AS(IEC)61511 term) - failure related in a deterministic way to a certain cause, which can only be eliminated by a modification of the design or of the manufacturing process, operational procedures, documentation or other relevant factors.

systematic failure (AS(IEC)62061 term) - failure related in a deterministic way to a certain cause, which can only be eliminated by a modification of the design or of the manufacturing process, operational procedures, documentation or other relevant factors.

systematic failure (ISO13849 term) - failure related in a deterministic way to a certain cause, which can only be eliminated by a modification of the design or of the manufacturing process, operational procedures, documentation or other relevant factors.

systematic safety integrity (AS(IEC)61508 term) - part of the safety integrity of a safety-related system relating to systematic failures in a dangerous mode of failure.

systematic safety integrity (AS(IEC)61511 term) - that part of the safety integrity of safety instrumented function relating to systematic failures in a dangerous mode of failure.

systematic safety integrity (AS(IEC)62061 term) - part of the safety integrity of a SRECS or its subsystems relating to its resistance to systematic failures in a dangerous mode.

target failure measure (AS(IEC)61508 term) - target probability of dangerous mode failures to be achieved in respect of the safety integrity requirements, specified in terms of either the average probability of a dangerous failure of the safety function on demand, (for a low demand mode of operation); the average frequency of a dangerous failure (for a high demand mode of operation or a continuous mode of operation).

target failure measure (AS(IEC)61511 term) - intended probability of dangerous mode failures to be achieved in respect of the safety integrity requirements, specified in terms of either the average probability of failure to perform the design function on demand (for a demand mode of operation) or the frequency of a dangerous failure to perform the SIF per hour (for a continuous mode of operation).

target failure value (AS(IEC)62061 term) - intended PFHD to be achieved to meet a specific safety integrity requirement(s).

target risk (AS(IEC)61508 term) - risk that is intended to be reached for a specific hazard taking into account the EUC risk together with the E/E/PE safety-related systems and the other risk reduction measures.

test harness (AS(IEC)61508 term) - facility that is capable of simulating (to some useful degree) the operating environment of software or hardware under development, by applying test cases to the software and recording the response.

test rate (r_t) (ISO13849 term) - frequency of automatic tests to detect faults in a SRP/CS, reciprocal value of diagnostic test interval.

tolerable risk (AS(IEC)61508 term) - risk which is accepted in a given context based on the current values of society.

tolerable risk (AS(IEC)61511 term) - risk which is accepted in a given context based on the current values of society.

trip device (AS4024.1 term) - device which causes a machine or machine elements to stop (or ensure an otherwise safe condition) when a person or a part of his body goes beyond a safe limit. Trip devices may be (a) mechanically actuated: e.g. trip wires, telescopic probes, pressure sensitive devices,etc.; or (b) non-mechanically actuated: e.g. photoelectric devices, devices using capacitive, ultrasonic or similar means to achieve detection.

two-hand control device (AS4024.1 term) - hold-to-run control device that requires at least the simultaneous actuation of two manual controls (actuators) in order to trigger and to main operation of the machine or machine elements, thus affording a measure of protection for the person operating the manual control (actuators).

type A standards (fundamental safety standards) (AS4024.1 term) - standards giving basic concepts, principles for design, and general aspects that can be applied to all machinery.

type B standards (group safety standards) (AS4024.1 term) - standards dealing with one safety aspect or one type of safety related device that can be used across a wide range of machinery: (a) Type B1 standards on particular safety aspects (e.g. safety distances, surface temperature, noise). (b) Type B2 standards on safety-related devices (e.g. two-hand controls, interlocking devices, pressure sensitive devices, guards).

type C standards (machine safety standards) (AS4024.1 term) - standards giving detailed safety requirements for a particular machine or group of machines.

uncontrolled stop (AS4024.1 term) - the stopping of machine motion by removing power to the machine actuators, all brakes and/or other mechanical stopping devices being activated.

undetected / unrevealed / covert (AS(IEC)61508 term) - in relation to hardware, undetected by the diagnostic tests, proof tests, operator intervention (for example physical inspection and manual tests), or through normal operation.

undetected / unrevealed / covert (AS(IEC)61511 term) - in relation to hardware and software faults not found by the diagnostic tests or during normal operation.

unexpected start-up/ unintended start-up (AS4024.1 term) - any start-up cased by (a) a start command which is the result of a failure in, or an external influence on, the control system; (b) a start command generated by inopportune action on a start control or other parts of the machine as, e.g. a sensor of a power control element; (c) restoration of the power supply after an interruption; or (d) external/internal influences (gravity, wind, self-ignition in internal combustion engines) on parts of the machine.

utility software (AS(IEC)61511 term) - software tools for the creation, modification, and documentation of application programs. These software tools are not required for the operation of the SIS.

validation (AS(IEC)61508 term) - confirmation by examination and provision of objective evidence that the particular requirements for a specific intended use are fulfilled.

validation (AS(IEC)61511 term) - activity of demonstrating that the safety instrumented function(s) and safety instrumented system(s) under consideration after installation meets in all respects the safety requirements specification.

validation (AS(IEC)62061 term) - confirmation by examination (e.g. tests, analysis) that the SRECS meets the functional safety requirements of the specific application.

verification (AS(IEC)61508 term) - confirmation by examination and provision of objective evidence that the requirements have been fulfilled.

verification (AS(IEC)61511 term) - activity of demonstrating for each phase of the relevant safety life cycle by analysis and/or tests, that, for specific inputs, the outputs meet in all respects the objectives and requirements set for the specific phase.

verification (AS(IEC)62061 term) - confirmation by examination (e.g. tests, analysis) that the SRECS, its subsystems or subsystem elements meet the requirements set by the relevant specification.

watchdog (AS(IEC)61511 term) - combination of diagnostics and an output device (typically a switch) for monitoring the correct operation of the programmable electronic (PE) device and taking action upon detection of an incorrect operation.

Appendix B: Project Safety Lifecycle Checklist

Safety Lifecycle Phase	Name of the Phase	Applicable Standard	Responsible Entity	Activities to be Undertaken	Complete	Documentation to be Kept	Complete
1	Concept	AS(IEC)61508	End User	A thorough familiarity should be acquired of the proposed EUC, its anticipated control functions and its physical environment.		Functional specifications, overview drawings and concept schematics of the EUC and its control system, if available.	
				The likely sources of hazard should be determined.		Previous risk assessments conducted by the EUC supplier and/or other users.	
				Information about hazard sources should be obtained (eg. toxicity, explosive conditions, electrical, corrosiveness, reactivity, flammability etc…).		Literature review of previous safety incident and accident data.	
				Information about the current safety regulations (national and international) pertaining to the type of EUC (eg. winder, conveyor, mobile plant etc…) should be obtained.		MSDS for hazardous materials.	
				Hazards due to interactions with other EUC's (already installed, or to be installed) in the proximity of the EUC should be considered.		Copies of relevant extracts from safety legislation, regulations, codes of practice and safety alerts pertaining to the type of EUC.	
						Proposed layout drawings and concept of operations for the EUC at the site, including its interactions with other plant / EUC's and with people (including operators, workers, maintainers and the public).	

Safety Lifecycle Phase	Name of the Phase	Applicable Standard	Responsible Entity	Activities to be Undertaken	Complete	Documentation to be Kept	Complete
2	Scope	AS(IEC)61508	End User	The physical environment, including the EUC and EUC control system to be included in the scope of the hazard and risk analysis should be specified.		Functional Safety Management Plan (FSMP). This plan should also refer to sub-ordinate plans as follows:	
				The physical equipment, including the EUC and the EUC control system to be included in the scope of the hazard and risk analysis should be specified.		Verification Plan,	
				The external events to be included in the hazard and risk analysis should be specified.		Validation Plan.	
				The sub-systems associated with the hazards should be specified.		Scope of the Hazard and Risk Analysis to be conducted on the EUC.	
				The type of accident-initiating events that need to be considered (eg. component failures, human error, dependent failures) should be specified.			

Safety Lifecycle Phase	Name of the Phase	Applicable Standard	Responsible Entity	Activities to be Undertaken	Complete	Documentation to be Kept	Complete
3	Hazard and Risk Analysis	AS(IEC)61508, AS (IEC)61511, ISO12100, ISO14121, AS4024.1	End User	A hazard and risk analysis should be conducted, based on the scope defined in Phase 2.		Hazard and Risk Analysis Report.	
				Consideration should be given to the elimination or reduction of hazards.		Documents used during the hazard and risk analysis (eg. corporate risk matrix, drawings and specifications etc.).	
				There should be consideration of the hazards that may occur under all 'reasonably foreseeable' circumstances (including fault conditions, misuse and malevolent / unauthorized action).			
				The event sequences leading to the hazards should be determined.			
				The likelihood of the hazardous events should be evaluated.			
				The potential consequences of the hazardous events should be determined.			
				The EUC risk (based on the combination of likelihood and consequence) should be evaluated for each determined hazardous event.			
				The hazard and risk analysis should also consider:			
				– The contributing components of each hazard.			
				– The consequences and likelihood of each hazardous event.			
				– The Necessary Risk Reduction for each hazardous event.			
				– The measures used to reduce or remove hazards and risks.			
				– The assumptions made during the analysis of the risks.			

In addition, I recommend that the following actions are taken to ensure that Phase 3 of the functional safety lifecycle is consistent with the requirements of the Model Workplace Health and Safety Act and Regulations 2011:

1. Risk elimination measures must be used against a hazard, where "reasonably practicable", before risk reduction measures are considered.
2. If risk elimination measures are not "reasonably practicable", the hierarchy of risk controls must be used to guide the selection of the most effective risk reduction measures for each hazard.
3. The selection of risk reduction measures for each hazard must encompass all those that are known, available and suitable at the particular time. All that can be done should be done.
4. If the potential severity of harm from the hazard is significant (ie. serious injury or death) then all known, available and suitable risk reduction measures should be implemented, regardless of cost, or the activity giving rise to the hazard should not be undertaken.
5. Safety requirements allocation against a hazard using a 'tolerable risk' approach should not occur until there is a level of confidence that the selection of risk reduction measures to be used against the hazard would allow the "reasonably practicable" test of the PCBU's primary duty to be satisfied.

Safety Lifecycle Phase	**Name of the Phase**	**Applicable Standard**	**Responsible Entity**	**Activities to be Undertaken**	**Complete**	**Documentation to be Kept**	**Complete**
4	Overall Safety Requirements	AS(IEC)61508, AS(IEC)61511	End User	The hazards, risk levels and risk controls from Phase 3 should be used as a basis for identification of safety functions and safety-related systems.		A Safety Requirements Specification (SRS) for the safety-related systems should be initiated during this phase, and should include, as appropriate:	
				A "tolerable risk" frequency should be established so that Safety Requirements Allocation may be performed during AS(IEC)61508 Phase 5.		– A confirmation of which safety functions are to be implemented (in whole or in part) via E/E/PE technologies.	
				If security threats have been identified then a 'vulnerability analysis' should be undertaken to specify security requirements (see IEC62443).		– A confirmation of which safety functions are to be implemented (in whole or in part) via 'other technologies'.	
						– A confirmation of the "other risk reduction measures" (ie. guards, barriers, procedures etc…) which are required to be implemented in conjunction with the safety-related systems.	
						– The expected environmental conditions and the general requirements on the safety-related systems.	
						– A definition of the "safe state" of the EUC.	
						– A definition and justification of the 'tolerable risk' target to be used during Phase 5.	
						Vulnerability Analysis Report, if required.	

Safety Lifecycle Phase	Name of the Phase	Applicable Standard	Responsible Entity	Activities to be Undertaken	Complete	Documentation to be Kept	Complete
5	Safety Requirements Allocation	AS(IEC)61508, AS(IEC)61511, AS(IEC)62061, ISO13849	End User	The hazards, risk levels and risk controls from Phase 3 should be used as an input to this step.		Safety Requirements Allocation Report (if Hazard and Risk Analysis and Safety Requirements Allocation are co-conducted, then a combined report may be provided),	
				The concept of "tolerable risk" from Phase 4 should be used as an input to this step (AS(IEC)61508 only).			
				The safety integrity requirements for each safety function (whether E/E/PE or "other technology" based) should be determined by a SIL or PL allocation conducted according to an appropriate method.			
				Where a safety function is to be implemented by E/E/PE technologies, then its functional and safety integrity requirements should be specified per AS(IEC)61508.1 Clause 7.10, AS(IEC)62061 Clause 5.2, AS(IEC)61511 Clause 10 or ISO13849 Clause 5, as appropriate.			
				Where a safety function is to be implemented by "other technology" (eg. pneumatic, hydraulic, mechanical etc…) then its functional and safety integrity requirements should be specified per ISO13849-1 Clause 5.			
				If a safety function is intended to be implemented by a combination of "other technology" and E/E/PE parts, then the functional and safety integrity requirements of the individual parts should be specified via a standard most appropriate to the technology. Table 1 of AS(IEC)62061 and Table 1 of ISO13849-1 provide guidance on the applicability of AS(IEC)62061 and ISO13849-1 to E/E/PE and "other technology" parts.			

Safety Lifecycle Phase	Name of the Phase	Applicable Standard	Responsible Entity	Activities to be Undertaken	Complete	Documentation to be Kept	Complete
				Where the proposed set of risk controls for a hazard is insufficient to achieve a "tolerable risk", an additional safety function must be considered.			
				The functional and safety integrity requirements of each safety function comprising the E/E/PE and "other technology" safety-related systems should be documented in a Safety Requirements Specification (SRS), which should form part of the overall specification of the EUC.			
				Any "other risk reduction measures" (ie. guards, barriers, procedures etc.) that are intended to work in co-ordination with the safety functions should be defined in the SRS.			

In addition, I recommend that the following action is taken to ensure that Phase 5 of the functional safety lifecycle is consistent with the requirements of the Model Workplace Health and Safety Act and Regulations 2011:

1. Safety requirements allocation against a hazard using a 'tolerable risk' approach should not occur until there is a level of confidence that the selection of risk reduction measures to be used against the hazard would allow the "reasonably practicable" test of the PCBU's primary duty to be satisfied.
2. The allocation of levels of safety integrity (ie. SIL, PL or CAT) to any safety-related systems must also take account of the level of safety integrity that would be "reasonably practicable" for a designer to achieve.
3. Where a regulation, code of practice or standard specifies a level of safety integrity or other specific requirements for a particular type of safety-related system, and this is more onerous than the SIL, PL or CAT allocated, then this requirement should take precedence.

A verification is recommended at the completion of Phase 5, covering the specific objectives, requirements, inputs and deliverables for Phases 1 to 5. A Verification Report should be produced.

Safety Lifecycle Phase	Name of the Phase	Applicable Standard	Responsible Entity	Activities to be Undertaken	Complete	Documentation to be Kept	Complete
9	Safety Requirements Specification	AS(IEC)61508, AS(IEC)61511, AS(IEC)62061, ISO13849	End User	The functional and safety integrity requirements of each safety function comprising the E/E/PE and "other technology" safety-related systems should be documented in a Safety Requirements Specification (SRS), which should form part of the overall specification of the EUC.		The Safety Requirements Specification (SRS) for the E/E/PE and "other technology" safety-related systems is completed during Phase 9, and should contain the information listed in AS(IEC)61508.1 Clause 7.10.2, AS(IEC)62061 Clause 5, AS(IEC)61511 Clause 10 or ISO13849-1 Clause 5 , as appropriate. For example, if AS(IEC62061 is to be used to develop the SRS, the following information should be provided:	
				Any "other risk reduction measures" (ie. guards, barriers, procedures etc.) that are intended to work in co-ordination with the safety functions should be defined in the SRS.		1. A confirmation of which Safety Functions are to be implemented (in whole or in part) via E/E/PE technologies.	
						2. A confirmation of which safety functions are to be implemented (in whole or in part) via "Other Technologies".	
						3. A confirmation of the "other risk reduction measures" (ie. guards, barriers, procedures) which are required to be implemented in conjunction with the safety-related systems.	
						4. The expected environmental conditions and the general requirements on the safety-related systems.	
						5. A definition of the "safe state" of the EUC.	
						6. A statement of the required performance of each safety function, via the following:	

Safety Lifecycle	Name of the Phase	Applicable	Responsible Entity	Activities to be Undertaken	Complete	Documentation to be Kept	Complete
						i) A name, identifier and description of each safety function,	
						ii) The required SIL or PL to be achieved by the safety function and the demand mode.	
						iii) A statement of the functional intent of each safety function.	
						iv) The frequency of operation / demands,	
						v) A description of the fault reaction required,	
						vi) The required response time of the safety function and its input / output devices,	
						vii) Interfaces to other systems or functions,	
						viii) The conditions or operating modes of the EUC under which the safety function will be active and disabled,	
						ix) The priority of those functions that can be simultaneously active, but can cause conflicting action in the EUC,	
						x) A description of the operating environment,	
						xi) Tests and associated facilities required,	
						xii) Rate of operating cycles, duty cycle, demand rate and utilisation.	

A verification is recommended at the completion of Phase 9, covering the specific objectives, requirements, inputs and deliverables for Phase 9. Particular attention should be paid to the safety Requirements Specification (SRS). A Verification Report should be produced.

Safety Lifecycle Phase	Name of the Phase	Applicable Standard	Responsible Entity	Activities to be Undertaken	Complete	Documentation to be Kept	Complete
6	Planning for Operations and Maintenance	AS(IEC)61508, AS(IEC)61511, AS(IEC)62061, ISO13849	End User	Develop a plan for operating and maintaining the E/E/PE and "other technology" safety-related systems and "other risk reduction measures".		Operations and Maintenance Plan, including:	
						1. The routine actions which need to be carried out on the E/E/PE safety-related systems, "other technology" safety-related systems and "other risk reduction measures",	
						2. The actions and constraints that are necessary to prevent an unsafe state, to reduce demands on the E/E/PE safety-related systems, "other technology" safety-related systems and "other risk reduction measures" or to reduce the consequences of hazardous events,	
						3. The documentation that needs to be maintained showing the results of functional safety audits and tests,	
						4. The documentation that needs to be maintained on hazardous incidents and all incidents with a potential to create a hazard event,	
						5. The scope of the maintenance activities (as distinct from modification activities),	
						6. The actions to be taken in the event of hazards occurring,	
						7. The contents to be included in the chronological documentation of the operations and maintenance activities ,	
						8. The requirement that the routine maintenance activities which are carried out to detect unrevealed faults should be determined by a systematic analysis, and	
						9. The requirement that the plan be agreed upon by those responsible for operating and maintaining the E/E/PE safety-related systems, "other technology" safety-related systems, "other risk reduction measures" and non-safety-related systems which have the potential to place demands upon the safety-related systems.	

Safety Lifecycle Phase	Name of the Phase	Applicable Standard	Responsible Entity	Activities to be Undertaken	Complete	Documentation to be Kept	Complete
7	Planning for Validation	AS(IEC)61508, AS(IEC)61511, AS(IEC)62061, ISO13849	End User	Develop a plan for validating the E/E/PE and "other technology" safety-related systems and "other risk reduction measures".		Validation Plan, including:	
						1. Details of when the validation shall take place,	
						2. Details of those who will perform the validation,	
						3. Specification of the relevant modes of operation of the EUC and their relationships to the E/E/PE safety-related systems, "other technology" safety-related systems and "other risk reduction measures",	
						4. Specification of the E/E/PE safety-related systems, "other technology" safety-related systems and "other risk reduction measures", to be validated for each mode of EUC operation (before commissioning commences),	
						5. The technical strategy for the validation (eg. analytical methods, tests etc...),	
						6. The measures, techniques and procedures that shall be used for confirming that the allocation of safety functions has been carried out correctly (confirmation shall ensure that each safety function conforms to its functional and safety integrity requirements),	
						7. Specific reference to the achievement (or non-achievement) of each requirement specified in the Safety Requirements Specification (SRS),	
						8. The required environment in which the validation tests will take place,	
						9. The pass and fail criteria,	
						10. The policies and procedures for evaluating the results of the validation, particularly non-compliances.	

Safety Lifecycle Phase	Name of the Phase	Applicable Standard	Responsible Entity	Activities to be Undertaken	Complete	Documentation to be Kept	Complete
8	Planning for Installation and Commissioning	AS(IEC)61508, AS(IEC)61511, AS(IEC)62061, ISO13849	Designer	Develop a plan for installing and commissioning the E/E/PE and "other technology" safety-related systems and "other risk reduction measures" in a controlled manner.		Installation and Commissioning Plan, including:	
						1. The installation schedule,	
						2. Those responsible for the different parts of the installation,	
						3. Procedures for the installation,	
						4. The sequence in which the various elements are to be integrated,	
						5. The criteria for declaring that all or parts of the E/E/PE safety-related systems ready for installation and for declaring installation activities complete,	
						6. Procedures for resolution of failures and incompatibilities,	
						7. The commissioning schedule,	
						8. Those responsible for the different parts of the commissioning,	
						9. Procedures for the commissioning,	
						10. The relationships to the different steps in the installation,	
						11. The relationships to the validation.	

A verification is recommended at the completion of Phase 8, covering the specific objectives, requirements, inputs and deliverables for Phases 6 to 8. This verification may need to be performed in parts (one part for each phase) and it should be scheduled so as to allow time for rectification of the design if deficiencies are found. A Verification Report should be produced.

Safety Lifecycle Phase	Name of the Phase	Applicable Standard	Responsible Entity	Activities to be Undertaken	Complete	Documentation to be Kept	Complete
10	Design of E/E/PE Safety-related Systems (using AS(IEC)62061)	AS(IEC)62061	Designer	Design and development of the E/E/PE safety-related systems for machinery applications should follow a ten (10) step process, as provided by Figure 2 of AS(IEC)62061. Guidance on the AS(IEC)62061 design and development process is provided in AS(IEC)62061 Clause 6.6.2 and Annex B.		Design documentation according to AS(IEC)62061 Annex B and Clause 10.	
				The key focus of the design must be to the following aspects:		Safety Integrity Level (SIL) Design Verification Report - PFH / PFD calculations, analysis of architectural constraints based on sub-system types, SFF and HWFT.	
				- Satisfaction of the architectural constraints for the applicable SIL for each SRECS sub-system involved in performing the SRCF's. This will usually involve the calculation or sourcing of Safe Failure Fraction (SFF) and dangerous failure rate values for the components / modules to be used,		Evidence of compliance to the requirements of AS(IEC)61508.2 AS(IEC)61508.3, or AS(IEC)62061 for avoidance and control of systematic failure (also see Sections 7.15 and 7.16).	
				– Achieving either a total Probability of Failure on Demand (PFD) or Probability of Failure per Hour (PFH) value for each SRCF, as applicable, that complies with the target failure values for the SIL that was allocated to the SCRF. This may involve an iterative reliability analysis to ensure that the detailed design and component selections are sufficient to achieve the target SIL.		Information for use according to AS(IEC)62061 Clause 7.	
				– Achieving the requirements for avoidance and control of systematic failures for both hardware and software used in the SRECS. These requirements are detailed in AS(IEC)62061 Clauses 6.7.9, 6.11 and Annex C, AS(IEC)61508.2 and AS(IEC)61508.3 Annex A. This will involve the gathering of evidence of activities undertaken during the specification, design and implementation stages in order to make a deterministic argument that the systematic failure requirements of the target SIL have been met.			

Safety Lifecycle Phase	Name of the Phase	Applicable Standard	Responsible Entity	Activities to be Undertaken	Complete	Documentation to be Kept	Complete
10	Design of E/E/PE Safety-related Systems (using ISO13849)	ISO13849	Designer	Design and development of the E/E/PE safety-related systems should follow the requirements of ISO13849-1 Clause 4 and Figure 3 for the allocated PL. Note: *It is the author's view that the approach to the specification and realisation of machinery-sector safety-related systems provided by the latest issues of ISO13849 be used, and the out-dated approaches provided by the 2006 issues of AS4024.1501 and AS4024.1502 should be avoided where possible.*		Sufficient documentation should be generated that describes the design of the "other technology" safety-related system and the achievement of its specific requirements. The list of 'information for use' in ISO13849-1 Clauses 10 and 11 provide a reasonable guideline for this.	
				The key focus of the design must be to the following aspects:		Design and verification documentation for each safety function, per ISO13849-1 Clause 4.	
				– Achieving the reliability and architectural requirements of the allocated PL for each safety function in terms of designated CAT architecture, DC_{ave}, CCF, $MTTF_d$ and PFH requirements. This may involve an iterative reliability analysis to ensure that the detailed design and component selections are sufficient to achieve the target PL.		Documentation for validation phase for each safety function, per ISO13849-2.	
				– Achieving the requirements for avoidance and control of systematic failures for both hardware and software used in the SRP/CS. These requirements are detailed in ISO13849-1 Clause 4.6 and Annex G. This will involve the gathering of evidence of activities undertaken during the specification, design and implementation stages in order to make a deterministic argument that the systematic failure requirements of the target PL have been met.			

Safety Lifecycle Phase	Name of the Phase	Applicable Standard	Responsible Entity	Activities to be Undertaken	Complete	Documentation to be Kept	Complete
11	Design of "Other Technology" Safety-related Systems	ISO13849	Designer	Design and development of the "other technology" safety-related systems should follow the requirements of ISO13849-1 Clause 4 and Figure 3 for the allocated PL. Note: *It is the author's view that the approach to the specification and realisation of machinery-sector safety-related systems provided by the latest issues of ISO13849 be used, and the out-dated approaches provided by the 2006 issues of AS4024.1501 and AS4024.1502 should be avoided where possible.*		Sufficient documentation should be generated that describes the design of the "other technology" safety-related system and the achievement of its specific requirements. The list of 'information for use' in ISO13849-1 Clauses 10 and 11 provide a reasonable guideline for this.	
				The key focus of the design must be to the following aspects:		Design and verification documentation for each safety function, per ISO13849-1 Clause 4.	
				– Achieving the reliability and architectural requirements of the allocated PL for each safety function in terms of designated CAT architecture, DC_{ave}, CCF, $MTTF_d$ and PFH requirements. This may involve an iterative reliability analysis to ensure that the detailed design and component selections are sufficient to achieve the target PL.		Documentation for validation phase for each safety function, per ISO13849-2.	
				– Achieving the requirements for avoidance and control of systematic failures for both hardware and software used in the SRP/CS. These requirements are detailed in ISO13849-1 Clause 4.6 and Annex G. This will involve the gathering of evidence of activities undertaken during the specification, design and implementation stages in order to make a deterministic argument that the systematic failure requirements of the target PL have been met.			

Safety Lifecycle Phase	Name of the Phase	Applicable Standard	Responsible Entity	Activities to be Undertaken	Complete	Documentation to be Kept	Complete
11	Design of Other Risk Reduction Measures	As appropriate to the application	Designer	Design guidelines for "other risk reduction measures" are not within the scope of AS(IEC)61508. Design guidelines need to be applied which are appropriate to type and the application of the measure.		Sufficient documentation should be generated that describes the design of each "other risk reduction measure" and the achievement of its specific requirements, eg. for fixed or movable guards - sufficient evidence to demonstrate compliance to AS4024.1 Part 1601.	
				For example, fixed and movable guards which are used on machinery may be designed to AS4024.1 Part 1601 and any interlocking devices to be used with guards may be designed to AS4024.1 Part 1602.			
				The appropriate standards and guidelines pertaining to the design of each type of "other risk reduction measures" should be identified and applied early in the realisation stage.			

A verification is recommended at the completion of Phase 11, covering the specific objectives, requirements, inputs and deliverables for Phases 10 - 11. This verification may need to be performed in parts (one part for each phase) and it should be scheduled so as to allow time for rectification of the design if deficiencies are found. A Verification Report should be produced.

In the event that AS(IEC)62061 is used for the design of the electrical / electronic safety-related systems (Phase 10), intermediate verifications are included at Step 5 and Step 8 of the 10-step design and development process (see Section 7.12).

In the event that ISO13849-1 is used for the design of the electrical / electronic safety-related systems (Phase 10), verification is included at Step 6 of the 7-step design and development process (see Section 7.13).

Safety Lifecycle	Name of the Phase	Applicable Standard	Responsible Entity	Activities to be Undertaken	Complete	Documentation to be Kept	Complete
12	Installation & Commissioning	AS(IEC)61508, AS(IEC)61511, AS(IEC)62061, ISO13849	Designer	Installation and commissioning must be carried out in accordance with the Installation and Commissioning Plan developed during Phase 8.		Installation and Commissioning Report, including	
						1. The installation activities and resolution of failures and incompatibilities,	
						2. The commissioning activities and resolution of failures and incompatibilities, along with references to failure reports.	

Safety Lifecycle	Name of the Phase	Applicable Standard	Responsible Entity	Activities to be Undertaken	Complete	Documentation to be Kept	Complete
13	Validation	AS(IEC)61508, AS(IEC)61511, AS(IEC)62061, ISO13849	End User	Validation should be carried out in accordance with the Validation Plan developed during Phase 7, in accordance with the Functional Safety Management Plan,		Identification of the version of the SRS under which the validation was carried out.	
				All equipment used for quantitative measurements as part of the validation activities should be calibrated against a specification traceable to a national standard or to a vendor specification,		Identification of the safety functions being validated (by test or analysis),	
				When discrepancies occur between actual and expected results, the analysis made and the decisions taken on whether to continue the validation or issue a change request and return to an earlier part of the validation should be documented.		Tools and equipment used with calibration data,	
						Documentation of the validation activities,	
						The results of the validation activities, including recorded discrepancies between expected and actual results,	
						Configuration identification of the item under test, the procedures applied and the test environment.	
						A statement of the validator's competence and independence.	

A verification is recommended at the completion of Phase 13, covering the specific objectives, requirements, inputs and deliverables for Phases 12 -13. A Verification Report should be produced.

If a functional safety assessment is to be conducted, then it is recommended that this occur prior to the commencement of operation of the EUC (Phase 14). This should constitute a high-level audit of the application of the safety lifecycle processes, the key activities undertaken, the inputs and outputs, and the contents of the Safety Lifecycle Dossier for Phases 1 - 13. A Functional Safety Assessment Report should be produced.

Safety Lifecycle Phase	Name of the Phase	Applicable Standard	Responsible Entity	Activities to be Undertaken	Complete	Documentation to be Kept	Complete
14	Operations and Maintenance	AS(IEC)61508, AS(IEC)61511	End User	Preparation of plans and procedures for maintaining the safety-related systems; including operation, maintenance and repair procedures for hardware and software.		Operation & Maintenance manuals.	
				Implementation of plans and procedures.		Register of engineering changes made to hardware and software.	
				Checking compliance to maintenance schedules.		Register of changes to technical documentation.	
				Maintaining documentation.		The results of any functional safety audits.	
				Periodic functional safety audits.		Safety-related systems failures and incidents.	
				Documenting any modifications that are made.		Root Cause Analysis investigations into failures and incidents.	
						Evidence of regular checking of maintenance compliance.	
						Evidence of compliance to the Functional Safety Management Plan, Configuration Management Plan and other associated plans and procedures via Quality Management audits.	

Ongoing verifications are recommended during Phase 14, covering the specific objectives, requirements, inputs and deliverables for the phase. A Verification Report should be produced on each occasion.

Safety Lifecycle Phase	Name of the Phase	Applicable Standard	Responsible Entity	Activities to be Undertaken	Complete	Documentation to be Kept	Complete
15	Modification and Retrofit	AS(IEC)61508, AS(IEC)61511	End User	The modification or retrofit should only be initiated by the issue of an authorised request, under procedures provided in the Configuration Management Plan,		A plan of the modification process.	
				The reason for the modification should be documented as well as the hazards affected and the details of the proposed change,		Documentation of the decision-making process and each decision affecting a safety-related system.	
				An impact analysis of the modification should be performed to establish the effect of the proposed changes on the functional safety of the EUC, EUC control system, safety-related systems or "other risk reduction measures",		Procedures for defining a unique baseline of each version of the safety-related systems.	
				The assessment should include a hazard and risk analysis sufficient to determine the necessary breadth and depth of subsequent activities.		Means for definition of all configuration items of a baseline.	
				All accepted modifications that have an effect on the EUC, EUC control system, safety-related systems or "other risk reduction measures" should initiate a return to the appropriate phase.		A chronological documentation (e.g. logbook) of the change request procedures including;	
				A complete action plan should be prepared and documented before carrying out any modifications.		1. Identified hazards which can be affected;	
				Details on the specific requirements of these configuration management activities will be included in the verification and validation plans.		2. Description of the change request (hardware and/or software);	
						3. Reason(s) for the change request,	
						4. Decision made (and authorization for each decision);	
						5. The impact analysis;	
						6. Re-verification (of each phase) and revalidation;	
						7. All documents affected by the change request activities;	
						8. All activities which were carried out during the change process and the persons/entities who were responsible for them;	

Safety Lifecycle Phase	Name of the Phase	Applicable Standard	Responsible Entity	Activities to be Undertaken	Complete	Documentation to be Kept	Complete
						9. Any other documents that are generated by the return to a previous safety lifecycle phase (and all subsequent phases).	
						Documentation of the following information to permit a subsequent audit:	
						1. Configuration status;	
						2. Release status;	
						3. The justification for and approval of all modifications;	
						4. The details of the modification.	

A verification is recommended at the completion of each modification and retrofit (Phase 15) covering the specific objectives, requirements, inputs and deliverables for the phase. A Verification Report should be produced on each occasion.

Safety Lifecycle Phase	Name of the Phase	Applicable Standard	Responsible Entity	Activities to be Undertaken	Complete	Documentation to be Kept	Complete
16	Decommissioning	AS(IEC)61508, AS(IEC)61511	End User (with sub-contractor input)	Prior to decommissioning / disposal, an Impact Analysis should be carried out and should include an assessment of the impact of the proposed activity on the functional safety of the EUC, EUC control system, Safety-related Systems or "other risk reduction measures".		Details of the Impact Analysis of decommissioning / disposal upon the functional safety of the EUC, EUC control system, safety-related systems or "other risk reduction measures".	
				The Impact Analysis should also consider adjacent plant and systems and their impact on the EUC, EUC control system, safety-related systems or "other risk reduction measures".		Decommissioning / Disposal Request.	
				The assessment should include a hazard and risk analysis sufficient to determine the necessary breadth and depth of subsequent functional safety activities.		Decommissioning Plan.	
				Decommissioning / disposal should only be initiated by the issue of an authorised request, which should be dependent on the results of the Impact Analysis.		Any other documents that are generated by the return to a previous safety lifecycle phase (and all subsequent phases).	
				Prior to decommissioning / disposal taking place, a plan should be prepared which should include procedures for the closing down of the EUC, EUC control system, safety-related systems or "other risk reduction measures" and their dismantling.			
				If decommissioning / disposal has an impact on the functional safety of the EUC, EUC control system, safety-related systems or "other risk reduction measures", it should initiate a return to the appropriate lifecycle phase in order to address those inadequacies identified.			

A verification is recommended at the completion of decommissioning (Phase 16) covering the specific objectives, requirements, inputs and deliverables for the phase. A Verification Report should be produced.

References

Standards

AS/NZS ISO31000-2009 Risk Management - Principles and guidelines (replaces AS/NZS4360)

AS/NZS ISO31010-2011 Risk Management – Risk assessment techniques (replaces AS/NZS3931)

AS4024.-2006 Safety of Machinery

AS61508:2011 Parts 0 to 7: Functional Safety of electrical / electronic / programmable electronic safety-related systems.

AS IEC61511:2004 Parts 1 to 3: Functional Safety – Safety Instrumented Systems for the Process Industry.

AS62061:2006 Safety of Machinery - Functional safety of safety-related electrical / electronic / programmable electronic control systems.

HB89-2012 Risk management – Guidelines on risk assessment techniques

ISO12100:2010: Safety of machinery - General principles for design – Risk assessment and risk reduction

ISO13849-1:2006 Safety of machinery - Safety-related Parts of Control Systems - Part 1: General Principles for Design.

ISO13849-2: 2012 Safety of machinery - Safety-related Parts of Control Systems - Part 2: Validation.

ISO14121.1:2008 Safety of machinery - Risk Assessment - Principles

ISO14121.2:2012 Safety of machinery - Risk Assessment - Practical guidance and examples of methods

Textbooks

Goble, W.M., *Control Systems Safety Evaluation & Reliability,* 2nd Edition.

Johnstone, R. and Tooma, M., "Work Health and Safety Regulation in Australia – The Model Act", 2012.

Smith, D.J., *Reliability, Maintainability & Risk.*

Smith and Simpson, *"Functional Safety – A straightforward guide to applying IEC61508 and related standards",* 2nd Edition.

Articles and Reports

Johnstone, R., Bluff, L., and Quinlan, M., *"Submission to the National OH&S Review (Public Submission 055)".*

Minerals Council of Australia, *"Safety & Health Performance Report of the Australian Minerals Industry 2006-2007".*

Minerals Council of Australia, *"Safety & Health Performance Report of the Australian Minerals Industry 2003-2004".*

Minerals Industry Safety and Health Centre (MISHC), *"Final Report to Queensland Resources Council (QRC) on underlying Causes of Fatalities and Significant Injuries in the Australian Mining Industry"*, March 2005.

Minerals Industry Safety and Health Centre (MIHSC), *"Fatal Injuries Causal Analysis"* (a sub-project of a research project for the Queensland Resources Council 2004/2005).

Robinson. F., *"A study of occupational health and safety practice in the construction of the Snowy Mountains Hydro-Electric Scheme"*, National Occupational Health and Safety Commission of Australia.

Safe Work Australia, *"Work-related Traumatic Injury Fatalities, Australia 2010-2011"*, December 2012.

Sammarco, John. J. *"A Complexity Assessment Methodology for Programmable Electronic Mining Systems"*, National Institute for Occupational Safety and Health (NIOSH); Pittsburgh, Pennsylvania.

Government Publications

Australian Government, *"Australian Safety and Compensation Council, Public Discussion Paper – Safety Requirements for the Design, Manufacture and Conformity Assessment of Plant"*, February 2009.

New South Wales Department of Planning, *Hazardous Industry Planning Advisory Paper No.3 - Risk Assessment,* Consultation Draft, July 2008.

New South Wales Department of Primary Industries, *EES008.1 / EES008.2 / EES008.3 / EES008.4 / EES008.5: Design of Powered Winding Systems.*

New South Wales Department of Primary Industries, *Legislation Update No.5 / 2007.*

New South Wales Department of Primary Industries, *MDG1010:1997 Risk Management Handbook for the Mining Industry.*

New South Wales Department of Primary Industries, *MDG1014:1997 Guideline to Reviewing a Risk Assessment of Mine Equipment and Operations.*

New South Wales Department of Primary Industries, *Safety Alerts: SA09-03, SA06-13, SA06-12, and SA06-01.*

New South Wales Workcover Authority, "*Guide to the New South Wales Occupational Health and Safety Regulation Amendment (Major Hazard Facilities)"* 2008.

Safe Work Australia, "Interpretive Guideline – The meaning of Reasonably practicable", 2011.

Royal Commission Report on the Pike River Coal Mine Tragedy, Volumes 1 and 2, October 2012.

Websites

http://www.dguv.de/ifa/en/pra/softwa/sistema/index.jsp

http://www.iec.ch/news_centre/release/nr2002/nr1402.htm

http://www.internationallawoffice.com/Newsletters/detail.aspx?g=d2a7ec89-fd1b-43ea-b8aa-fc46fec4cbcf

http://www.itl.nist.gov/div898/handbook/apr/section4/apr451.htm

http://www.marcuspunch.com/index_files/fast.htm

http://en.wikipedia.org/wiki/Great_Wall_of_China

Index

Made in United States
North Haven, CT
10 November 2026

22095617R00135